A Guide to SPSS/PC+

Neil Frude

A Guide to SPSS/PC+

Second Edition

Springer-Verlag
New York Berlin Heidelberg Paris
Tokyo Hong Kong Barcelona Budapest

First published 1987 by
THE MACMILLAN PRESS LTD
London and Basingstoke

Reprinted 1989, 1990, 1992
Second edition 1993

Sole distributors in the United States of America, Canada and Mexico
SPRINGER-VERLAG NEW YORK INC.
175 Fifth Avenue,
New York, NY 10010

Printed in Hong Kong

ISBN 0–387–91442–0 Springer-Verlag New York

Library of Congress Cataloging-in-Publication Data is available.

Contents

Preface

SPSS/PC+ (the PC version of the Statistical Package for the Social Sciences) is the best-selling statistical analysis program for microcomputers. It enables data from surveys and experiments to be analyzed fully and flexibly. It has facilities for the extensive manipulation and transformation of data, and includes a wide range of procedures for both simple and highly complex statistical analysis. It also provides the opportunity for the researcher to produce fully labelled tables and graphs which can be easily incorporated into a final project report

Over the 25 years since it was first devised, the versatile SPSS system has become an indispensable tool for many workers in social science research (including psychology, sociology, politics, human geography, business management, etc.) and in business and government. Many of the largest and most important surveys in the past quarter century have been analyzed using one or other version of the system. SPSS (including the microcomputer version) is regularly used by government agencies, and by many major industrial corporations, market research companies and opinion poll organizations.

Familiarity with SPSS/PC+ will do more than merely enable students to analyze their current projects. Hands-on experience with the SPSS system will help them to develop valuable statistical skills and will provide a working knowledge of a system which will certainly continue to play a central role in social science research, and many commercial fields, for the foreseeable future. When you have worked through this book you can feel confident about having a basic knowledge of a package which is the best of its kind, and the skills you will have gained are likely to remain useful in a range of contexts, and however advanced your project work becomes.

Unfortunately, there is a problem. Although SPSS/PC+ is not difficult to *use*, many beginners find it difficult to *learn to use* the package. Although the documentation for SPSS/PC+ is excellent in many respects, the encyclopaedic nature of the manuals can prove a major obstacle to those wishing to become acquainted with the system for the first time. While the standard SPSS/PC+ manuals are indispensable as reference texts for the experienced user, they are not ideal for beginners who wish to learn to use the system.

There are several reasons for regretting the fact that beginners often experience difficulties in getting acquainted with the SPSS system, and that teachers often prefer to introduce the student to relevant areas of data analysis by employing less sophisticated packages:

- The power and flexibility of SPSS/PC+ means that the analysis of even modest datasets can be considerably enhanced through its use
- Experience with SPSS/PC+ provides students with an opportunity to explore the use of advanced multivariate techniques employed in many of the major studies about which they read
- SPSS has become a standard package within the social sciences and many related fields, so that students experienced in its use can feel confident that they have added significantly to their range of research- and employment-related skills.

This book aims to teach the use of SPSS/PC+ as clearly and as simply as possible. It is a stand-alone text covering many of the basic SPSS/PC+ procedures and may be used by the student working independently or may form the basis for class teaching. In either case it is assumed that the student has regular access to an SPSS/PC+ facility and will be able to work through parts of the text while interacting with the microcomputer.

Although this book does not assume a prior knowledge of survey work, statistics, computer programming or report writing, it is clear that one introductory book cannot teach all of these skills. Although this book describes aspects of survey methodology (particularly questionnaire design), students who need basic information about statistical tests may need to consult an introductory statistics textbook while working through this book. *A Guide to SPSS/PC+* is not concerned with computer programming (although it is concerned with computer *use*). SPSS/PC+ does not require the user to write programs. It is a computer program which enables the user to enter data, to describe, label and manipulate the data, and to perform simple and complex analyses. The various facilities offered by the program are activated by issuing commands that request procedures to be operated on specified data. The commands consist largely of keywords combined together following simple rules of syntax.

In writing this book I have placed the emphasis on simplicity and clarity rather than on exploring the full complexity of the system. When you have worked through this book (with hands-on experience of SPSS/PC+, using either your own data or that supplied in Appendix E) you will be well placed to cope with and benefit from the information provided in the SPSS/PC+ manuals. This book is not intended to replace these manuals and they remain the ultimate reference. For simple analyses, however, this book can act as a stand-alone text.

Unlike the SPSS/PC+ manuals, this book provides a progressive tutorial. The simplest and most fundamental concepts are introduced first. Later sections build on the knowledge which has been acquired at earlier stages. After a brief overview of the SPSS system, the reader is taught how to use a microcomputer to explore SPSS/PC+. Experience has shown that students often find the various types of 'file' used by SPSS/PC+ a source of some confusion, and special care has therefore been taken to distinguish carefully between the different types of file. These are introduced progressively through the text. This book focuses on data collected in a questionnaire survey, although SPSS/PC+ can also be used to analyze data from other types of study, including experiments. (Incidentally, although the word 'data' is sometimes thought of as singular, it is in fact plural, as in 'the data were collected' – *not* 'the data was collected'; a single data item is a 'datum'.)

A Guide to SPSS/PC+ provides step-by-step instructions for collecting information from a survey and then using SPSS/PC+ to analyze the data.

Although the book focuses particularly on data obtained from a questionnaire study, SPSS/PC+ can also be used to analyze data from other types of study, including epidemiological research and laboratory work. Whichever of these types of study is involved, the process leading up to the writing of a report can be thought of as comprising four stages. As you work through this book you will be given instructions on how to:

- collect raw data and code them into numerical form
- enter numerical data into the computer and save them
- define these data and annotate them with descriptions and labels
- use the dataset to produce summaries and tables, and to perform statistical tests

The chapters are arranged in a 'tutorial sequence', but after you have worked through the book once, you are likely to refer back to particular chapters when you wish to perform particular statistical analyses or to transform your data in various ways.

Chapter 1 provides a very concise overview of SPSS/PC+.

Chapter 2 shows how the data which will form the raw material for SPSS/PC+ analysis (using the example of a questionnaire survey) can be collected and coded.

Chapter 3 introduces the SPSS/PC+ system as it is implemented on a microcomputer. Details are provided of how the system can be accessed and how the user may switch between levels of program.

Chapter 4 takes the reader, step-by-step, through the procedures involved in using the SPSS/PC+ editor, REVIEW, to enter data into a computer file.

Chapter 5 shows how a data definition file may be written. This provides definitions and descriptors so that the dataset is fully specified to the system.

Chapter 6 discusses in detail two of the many statistical procedures which form the heart of the SPSS/PC+ system. These simple procedures (FRE-

QUENCIES and CROSSTABS) allow the user to examine some basic features of the dataset.

Chapter 7 introduces two further statistical procedures, MEANS and T-TEST.

Chapter 8 shows the reader how data can be transformed, re-ordered and selected for analysis.

Chapter 9 examines two further types of file which can be used when conducting SPSS/PC+ analysis.

Chapter 10 provides a further exploration of the editing and other facilities offered by the editor program, REVIEW.

Chapter 11 provides details of another statistical procedure, CORRELATION, and introduces PLOT, a procedure which can be used to plot frequency data in graphical form.

Chapter 12 adds analysis of variance to the range of analyses which the user can perform with SPSS/PC+.

Chapter 13 provides a detailed guide showing how SPSS/PC+ can be used to perform a wide range of non-parametric statistical procedures using the commands NPAR TESTS and RANK.

Chapter 14 examines the concept of the reliability of a scale and describes how the RELIABILITY procedure can be used to examine characteristics of a scale.

Chapter 15 describes how two or more related files can be joined together using the JOIN command.

Chapter 16 describes alternative ways of coding and entering data. It also discusses how 'portable files' can be created and transported from one computer to another. Finally, there are brief descriptions of many aspects of the SPSS/PC+ system which are beyond the scope of this introductory book.

A number of appendices are also provided, and a guide to these is presented on pp. xiv–xvi.

One way to use this book most effectively would be to devise a small study of your own – to construct a questionnaire and carry out a pilot survey (following Chapters 1 and 2), to create a **data file** and an appropriate SPSS/PC+ **data definition file** (following Chapters 4 and 5) and then to proceed to a simple analysis of your data (Chapters 6 and 7) before exploring some of the more complex aspects of the system (Chapters 8 to 10). The later chapters introduce a range of further statistical procedures and describe ways in which the system can be used with optimal efficiency and flexibility. When you have worked through this book you will be in a good position to explore more complex forms of SPSS/PC+ analysis with the aid of the SPSS/PC+ manuals. As an alternative to performing your own survey you might choose to make use of the data from the survey which was performed to provide many of the worked examples in this book. This survey concerns gender attitudes and was conducted on British psychology students. The questionnaire was specially constructed so that the data might allow various aspects of the SPSS/PC+ system to be explored. The original questionnaire is reproduced as *Appendix C*, and other appendices include the complete 'codebook' (used to code the data) and the data definition file used to make the data 'meaningful' to the system. The *complete* set of data is also reproduced so that students may, if they wish, work through many of the examples given in the book, following the various procedures on their own computer and exploring new ways of analyzing the data provided.

A Guide to the Appendices

Appendix A: Notes for Trainers

This appendix is intended for those who are involved in teaching others to use SPSS/PC+. The particular approach followed in this book is briefly explained, and there is some discussion about how the computer should be set up for the maximum ease of use by students who follow the instructions in the early chapters. The initial confusion which so often causes the beginner bewilderment and anxiety can be avoided if the trainer sets up the system to respond in ways that the student is led to expect, and if the trainer's teaching follows the strategies recommended in the text.

Appendix B: Using Word-processor and Other Files with SPSS/PC+

The use of the SPSS/PC+ editor, REVIEW, is discussed in a number of chapters of this book, and its use is recommended. However, it is sometimes more convenient to use another editing system (for example, a word-processing program) to produce files to be used by SPSS/PC+. This appendix describes how files produced by 'foreign' editor and word-processor systems can be made compatible with SPSS/PC+. Such files can later be re-edited using REVIEW. In addition, files produced by a wide range of database and spreadsheet programs can also be translated into SPSS/PC+ systems files.

The system automatically produces a listing file during each SPSS/PC+ session. This contains a copy of all of the tables, statistics, and other output produced during the session. *Appendix B* describes how this file can be edited with a word-processor.

Appendix C: The Gender Attitudes Survey: Questionnaire

This appendix introduces the 'Gender Attitudes Survey' and includes the original questionnaire used in the study.

Appendix D: The Gender Attitudes Survey: CODEBOOK

This reproduces the **codebook** which was produced to provide a guide to the **variables** represented within the Gender Attitudes Survey Questionnaire.

Appendix E: The Gender Attitudes Survey: DATA

This contains the full set of data collected from 50 (real!) respondents.

Appendix F: The Gender Attitudes Survey: DATA DEFINITION FILE

This appendix provides a full listing of the **data definition file** which was used to make the Gender Attitudes Survey data file accessible to SPSS/PC+.

Thus the four appendices, *Appendix C* to *Appendix F* constitute a complete package of materials relating to the Gender Attitudes Survey. Data from this small study are used throughout the book to illustrate the use of many different procedures. You might wish to create your own copy of the Gender Attitudes Survey dataset (by creating a copy of the data file and the data definition file) so that you can reproduce various analyses described throughout the text. You could also conduct your own original analyses of the dataset. However, do not attempt to create the relevant computer files until you have read the relevant chapters (Chapter 1 to Chapter 6) in this book.

If you are conducting your own survey you may wish to consult various aspects of the Gender Attitudes Survey package as a model for your own work.

Appendix G: Glossary of Statistical, Computing and SPSS/PC+ Terms

This provides a comprehensive glossary of the terms used in this book. It includes explanations of statistical terms, relevant computing terms and words which have a special function when used in SPSS/PC+. Whenever you encounter a word with which you are unfamiliar, or are unsure about the meaning of a term, you might find the help you need by consulting this Appendix. (If you are working within REVIEW, an 'on-line' glossary facility is also available – *see* Chapter 10.)

Appendix H: Types of File Used or Created by SPSS/PC+

This is a table that provides a brief description of features of the various types of **file** which you will encounter in using SPSS/PC+.

Appendix I: Variations between Different Versions of SPSS/PC+

If you are using a version of SPSS/PC+ other than the one specifically represented in this book (Version 4.0), you may wish to know how your version differs. This appendix provides an overview of some of the developments in SPSS/PC+ as it has evolved through successive versions. The main operating characteristics of the system have remained much the same through successive versions and are likely to remain similar in future versions. As the system has evolved, however, various enhancements have been added. These affect both the base module and the optional modules.

Appendix J: Brief guide to SPSS/PC+ Commands and Subcommands

This provides a brief synopsis of the SPSS/PC+ commands that are covered in detail elsewhere in this book. It will provide you with a handy reminder of the syntax of SPSS/PC+ commands, and of some of the subcommand options available within particular commands. It should be used only after you have become familiar with the various procedures and commands by reading through the relevant material in the main text.

Appendix K: Brief Guide to REVIEW Commands

This is another 'handy reminder' table, showing some of the key commands available when working within the SPSS/PC+ editor, REVIEW.

Acknowledgements

From the first edition

Many people, including students, colleagues, research workers and computer advisors have helped me in the task of creating this book. I approached some of them because of their expertise. Others were selected as 'naïve subjects' and introduced to SPSS for the first time. They worked through early drafts of the book, sitting before an Olivetti SP microcomputer, and made sure that what I said would happen *did* happen. Sometimes it did not, more often because of my mistake than theirs. Students were also asked to comment on difficulties they encountered in understanding the text, and some obviously relished the opportunity for role reversal which this offered. I amended the text in response to each critical note (including 'could be worded more elegantly' and 'are you sure you mean this?').

The 'experts' found bugs of a different species, and suitable pest control measures have been taken to try to eliminate these.

I am most grateful to all those who gave their time so generously. The principal helpers are listed below in alphabetical order. I will leave it to these individuals to judge whether I approached them for expert or 'grass roots' advice:

Annette Davies, Hilary Griggs, Rosemary Jenkins, Pamela Kenealy, Ray Kingdon, Mark McDermott and Martin Read

For the second edition

Preparation of the second edition was made easy by the fact that I have had the opportunity, over the five years since this book was first published, to help many readers (my own students, and the occasional correspondent) as they used the book to guide them through the analysis of their own projects. They, in turn, have helped me to discover the occasional inaccuracy and to recognize which aspects demanded clearer presentation. These students and correspondents are too numerous to mention individually, but I am grateful to them all.

1 Introducing SPSS/PC+

This chapter will lead you gently through the four stages involved in conducting an SPSS analysis. It provides an overview, and few details will be given at this stage. It is *not* important that you try to learn the particulars of the many aspects covered, but it *is* important that you read with understanding and gain a general impression of the SPSS approach.

The four stages which will be briefly covered involve:

- collecting raw data and coding them into numerical form
- entering numerical data into the computer and saving them
- defining these data and annotating them with descriptions and labels
- using the dataset to produce summaries and tables, and to perform statistical tests

1.1 Collecting and coding data

To examine facts relevant to social and psychological issues, social scientists need information, or data. Some data can be gathered from published sources, but a social scientist often has to collect new data by conducting an original study. This must be specially devised to provide the desired information or to answer specific questions.

The basic atom with which the social scientist deals is the **variable**. This can be any type of information item, including sex (i.e. male or female), age, income, marital status, the number of children a person has, an attitude, a belief, a personality score, the party a person voted for at the last election, etc.

One common type of study used in the social sciences is the **experiment**. Here one variable (e.g. the length of a list of words to be learned) is deliberately manipulated and the effect of this on another variable (e.g. the time needed to learn the list) is measured.

Another type of study is the **survey**. Often this will involve inviting people to participate by divulging information about themselves, their attitudes and their beliefs. The subjects (or **respondents**) may be interviewed or given questionnaires to complete. Certain characteristics of the respondents are noted (e.g. sex, age) and comprise some of the variables. The answers given to the questions are the other variables.

Some **questionnaires** may be standard tests (like personality tests) whereas others will be a series of questions put together by the researcher especially for the study.

Some **interviews** are relatively **unstructured** – more like a normal conversation – but others are **structured** – they closely follow a script (and are thus the verbal equivalent of a written questionnaire). The main advantage of unstructured interviews is that they yield more individual profiles of the respondent and allow interesting points to be probed in depth. The main advantage of structured interviews (or questionnaires) is that they allow the researcher to compare all the respondents' answers to the same questions given in the same order.

Within interviews or questionnaires, some questions permit respondents to answer in a free, or open, way. Thus:

> **'What are your feelings about the way in which this country is currently governed?' ....**

The main advantage of **open-ended questions** like this is that they allow a range of individual expression and thus provide rich and varied data. Unfortunately, however, open-ended questions are not easy to analyze. It is difficult to compare the answers which different respondents give to the same question.

Other questions (**closed questions**) are supplied with a fixed range of permitted responses. Here are three examples of closed questions:

(i) 'Do you believe in the existence of God?' YES NO

(ii) 'If you had to choose a painting for your room would you choose:
a) a fine print of an Old Master
b) an original 19th century water-colour
c) a modern painting?'

(iii) 'Indicate, by circling ONE number, where you 'stand politically' on the 'left-right' dimension:

Extreme left						**Extreme right**
1	**2**	**3**	**4**	**5**	**6**	**7'**

The main advantage of closed-ended questions is that all respondents answer within the same restricted range, thus making comparisons between respondents very easy.

For data to be used by SPSS they must be coded for all subjects in the same way. For each respondent the **value** for any variable (for example, the answer to a question) must be determined. Thus the variable *sex* has two possible valid values (corresponding to 'male' and 'female'). A coding scheme must be devised so that one **code** (say, 1) corresponds to 'male', and another code (say, 2) corresponds to 'female'. A variable such as *age in years* hardly needs a coding system because the form in which the data are collected (23, 56, etc.) can stand. Interviews (or written questionnaires) with closed-ended questions are easy to code because each of the range of permitted answers is assigned a particular code value (perhaps 1 for 'yes', 2 for 'don't know' and 3 for 'no'). Closed-ended questions thus yield information in a form which is ideal for SPSS analysis. Open-ended questions are likely to raise problems, however. Although it may be possible to devise a workable method of coding the data obtained from such questions, this can prove difficult in practice.

Applying the coding system devised by the researcher, all of the raw data from a survey (including the subject's identification number, sex, social class, age and the responses to particular questionnaire items) are transformed into numbers (letters can also be used, but we will not discuss such 'alphanumeric' variables until Chapter 16). The data for each respondent are thus represented as a series of numbers, and the data from all of the respondents together form the **dataset** on which SPSS/PC+ will operate.

The dataset therefore consists of data from a number of respondents (or cases) and the data for each case are represented by a series of numbers. These numbers represent the coded **values** (for the particular respondent) of data items relating to a number of different **variables**. Transforming raw data into values is achieved by applying the coding system.

1.2 Entering and saving data

Once all the data obtained from a study have been collected and coded into numerical form they are entered into the computer as a set of numbers. This consists of a table or 'matrix' of numbers, like the following matrix (but usually a lot bigger):

```
019865976597659
029867596576586
035475486547654
047865967654445
```

Each row (there are four rows in the above example) contains the data from a single respondent. The numbers in the columns along the rows (there are 15 columns in the example) represent the values which have been determined for each subject for particular variables (obtained from, say, the answers to a questionnaire). The whole **matrix** represents the **dataset**.

To prepare the data for entry into the computer we use a program referred to as an **editor**. We type the data across the screen as a long series of numbers. Each time we have finished typing the data for one respondent we start a new line (row) of data by pressing <RETURN>.

When the complete dataset has been 'keyed in' we follow instructions to save the data and the machine then stores the data matrix as a **data file** with a particular name. It is permanently stored on a magnetic **disk**.

We can call up the dataset at any time to use it, to modify it or to add to it. Once our dataset has been established we never need to type it in again. Thus we can perform many separate analyses, on many different occasions, after having typed in our data matrix only once. This facility for the 'once and for all' entry of our data is one of the great advantages of working with a 'package' such as SPSS/PC+ rather than with separate programs for analysis.

1.3 Annotating data

So far the data has been collected, coded and entered into the machine as a data matrix. But no instructions have been given to the machine about how to interpret or label the numbers. The next stage is to describe the data so that the system can identify the variables represented within the dataset.

The first thing we do is to tell the machine how to 'slice' the long string of numbers for any one subject, thus separating out the variables. *We* know (but the system doesn't) that the first row of numbers:

```
019865976597659
```

represents not *one* 15-digit number but rather a *series* of smaller numbers.

The 15-digit string *could* represent 15 single-digit numbers:

```
0/1/9/8/6/5/9/7/6/5/9/7/6/5/9
```

or it *could* represent a 'mixed bag' of single-, double- and triple-digit numbers:

```
01/9/8/65/9/76/5/9/7/659
```

i.e.

```
01,   9,   8,   65,   9,   76,   5,   9,   7,   659
```

In this case the 15 **digits** would represent coded values for only 10 **variables**.

The instructions given to the computer on how to slice up the dataset are called **format instructions**. The most usual way to arrange a dataset is in **fixed format**, i.e. where the data for all respondents has precisely the same format. Thus, for *all* respondents represented in the dataset, the data relating to a particular variable occupy precisely the same position (the same column or columns) along the long row of numbers. Each variable within the dataset is therefore represented by a one-, two- or three-digit (etc.) vertical slice through the data matrix.

Thus for all the data in the matrix displayed above, the arrangement in terms of variables (let's call them *V1* to *V10*) would be:

```
01   9   8   65   9   76   5   9   7   659
02   9   8   67   5   96   5   7   6   586
03   5   4   75   4   86   5   4   7   654
04   7   8   65   9   67   6   5   4   445

V1   V2  V3  V4   V5  V6   V7  V8  V9  V10
```

Obviously, the computer will need to be informed of the formatting arrangement for the dataset in order to locate variables within it. Since the same format is used for the data relating to all respondents (cases) within the dataset, however, a single format specification will be sufficient to cover all cases. Having been provided with the format specification, the computer will be able to recognise the numerical organization within the long string of digits which comprises the coded data for each respondent.

(In contrast to this fixed format system, SPSS/PC+ permits an alternative **free format** system. This is used less frequently, however, and need not be discussed at this stage. An explanation will be found in Chapter 16).

As we specify the format, we also invent a unique name for each of the variables, naming them – together with their position – in the order that they occur within the dataset. For some variables there will be an obviously appropriate name – thus for the sex of the respondent we are most likely to choose *sex* as the variable name. Other obvious names include *age*, *class* and *id* (for 'identification number').

In other cases there will not be an obvious choice of name. Suppose, for example, that a variable is the answer to the question: 'Do you believe in the existence of God'? We might name the variable *belief*, *beliefg*, *god*, *bgod* or *bg*. Names must be no more than eight characters in length, and spaces are not permitted, so *belief in god* would not be permissible as a variable name.

Although the best name to choose is often the one which will prove easiest to remember and recognize, it is sometimes difficult to capture the nature of the variable in a simple descriptive sequence of letters. It is often more manageable to use such names as *Q14* (meaning 'the coded answer to Question 14') or *VAR025* (meaning, simply, 'variable 25').

Once they have been established, variable names and format specifications will rarely need to be changed. Whatever analyses and manipulations will be performed with the dataset, the same variables will occupy the same locations within that dataset. Thus a statement which names the variables and specifies their format can become a permanent descriptor of the set of data.

So far, the system has been instructed about how the series of digits in the dataset are to be broken up into slices, each representing a variable, and it has been provided with names for these variables (in the order that they occur within the dataset). SPSS/PC+ also enables us to provide **variable labels** for each of the variables. Such labelling is entirely optional (data can be manipulated, tabulated and statistically analyzed without any labelling having been provided) but labels can annotate the computer output in a useful way. Screen displays and printed output are more easily understandable if they include labels. If the system displayed only the names of the variables analyzed (*var025*, *var034*, *bgod*, etc.) we would probably have to look back at our notes to identify particular variables. Output which is annotated with fuller labels, however, is easier to make sense of 'at a glance'.

Thus the system can be informed that *var025* should be labelled 'Number of church attendances in past year' or that *bgod* should be labelled 'Belief in the existence of God'.

SPSS/PC+ also allows us to provide **value labels** to label each of the values associated with any variable. Remember that data corresponding to each variable will have been coded and that each variable will have a range of valid values. Thus the raw data for the variable *soclass* (a label for this variable might be 'Social class') might have been coded into five categories. Each respondent for whom we have the raw data on social class will have been given a coded value (perhaps using the range 1 to 5) for the *soclass* variable. Particular values for this variable (1, 2, etc.) can now be labelled (e.g. 1 might be given the label 'Manual working class'). The two valid values for the variable *sex* could also be labelled (for example, the value 1 might be labelled 'Male' and the value 2 might be labelled 'Female'). Other commonly used value labels include 'Agree', 'Disagree', 'Yes', 'No', 'Always', 'Often', 'Sometimes' and 'Never'. Value labels, like variable labels, are entirely optional and are not used to communicate with or to instruct the system. They are included merely to enable the system to annotate the output as it is displayed on the screen or printed on paper. They are not necessary for the execution of any analysis but may add to the presentation and ease the interpretation of the output.

Everything that has been described in this section can be considered as permanently describing the dataset. Information about the format (or 'slicing') specifications, the variable names, the variable labels and the value labels can be stored permanently along with the data and will be used by the computer whenever we use SPSS/PC+ to analyse the particular dataset. Such information, therefore, need be entered only once and will remain in effect unless changes are deliberately made.

1.4 Using the annotated data

We are now at the most interesting and productive stage. It will soon be time for SPSS/PC+ to work for us. Some of the tasks it performs are relatively simple (but would be extremely tedious to perform by hand on a large dataset) while others are extremely complex.

Instructions to the system are given by various commands. Some of these (**operation commands**) affect the way in which the system works. Some other commands (**data definition** and **transformation commands**) identify or label existing variables, or create new variables, and some (**procedure commands**) request that analyses be performed on the data. Although commands are generally very easy to use, there are certain rules of syntax which need to be adhered to. These will be introduced progressively throughout this book.

One example of a **transformation command** is **COMPUTE**. This is used to create new variables from those which are already present in the dataset (by means of a computational action which may involve, for example, addition or multiplication). Suppose, for example, that our dataset consists of students' percentage scores on a number of class tests and their results in a final examination. We will name the variables *test1*, *test2*, *test3* and *exam*. Suppose that a student's final grade is obtained in the following way. Each of the three test scores (in percentage terms) are equally weighted and together contribute 50% of the final grade, with the examination mark (also expressed in percentage terms) contributing the other 50%.

SPSS/PC+ can be asked to compute a new variable, *grade*, using the existing values of the four variables for each student:

```
COMPUTE GRADE= (TEST1 + TEST2 + TEST3 + (EXAM*3)) / 6
```

This will yield, for each student, a value for a new variable *grade* (notice in the above example that * means 'multiply' and / means 'divide').

The above command instructs the computer to add each student's three test scores, then to add his or her exam score multiplied by 3, and then to divide the result of this computation by 6. The new variable *grade* will thus have a value for each student (equivalent to the percentage score for the final grade).

A **procedure command**, **FREQUENCIES**, can be used to instruct the system to return the frequency of cases (in the above example each student represents a single case) which have each of the values of a particular variable. Thus it will tell us how many students have a *grade* score of 45, how many have a score of 50, etc. The **FREQUENCIES** command can also be used to provide a number of other basic facts about the dataset. Thus it can be used to find the average of any variable (including both original variables in the dataset – *test1*, *test2*, etc. – and new computed variables – e.g. *grade*) or the **range** of any variable (i.e. the difference between the lowest and the highest values for the population of cases in the sample as a whole). **FREQUENCIES** can also be used to provide a graphical display of frequencies.

Suppose we want to obtain basic information about the variable *age* (e.g. how many 23-year olds there are in the sample, how many 25-year olds, the average age of the sample, etc.). The instruction **FREQUENCIES AGE.** would be sufficient to obtain the average (**mean**) age of our sample, an indication of the general spread (**standard deviation**) of the ages, and the minimum and maximum age (i.e. the ages of the youngest and the oldest) of our sample. The output will provide the frequency of each age and, if requested, supply a **histogram** or a **barchart** to illustrate this **frequency distribution**.

(Some of the above terms may have little meaning for you at this stage. Statistical and other terms are explained in the Glossary (*Appendix G*). Although no statistical knowledge is required to understand the directions given in this book for using SPSS/PC+, any proper use of the statistical procedures available in the package will of course require such knowledge. Consult one of the many statistics textbooks available.

After obtaining an overview of the average and spread of each single variable the researcher will generally wish to examine the relationship between pairs (or groups) of variables.

Thus it may be required to examine the relationship between the variables *sex*, ('male' and 'female') and *class* (say, 'working class' and 'middle class') to see whether, within the sample of respondents, the two sexes are unequally distributed between the social classes. An unequal distribution would mean that, for our sample, there was an association between *sex* and *class*. If we were to examine this question without the use of a computer package such as SPSS/PC+ we would need to count the number of:

```
working class males    working class females
middle class males     middle class females
```

It would then be useful to arrange for a display in the form of a 2 X 2 table with four 'cells', for example:

```
                                   SEX

                            Male        Female
                         .....................
                         .         .         .       Total W.C.
C      Working Class     .   45    .   `23   .         68
L                        .         .         .
A                        .....................
S                        .         .         .       Total M.C
S      Middle Class      .   19    .   38    .         57
                         .         .         .
                         .....................

                       Tot. Male   Tot. Female     TOTAL
                            64         61            125
```

Such a table is referred to as a 'crosstabulation' or a 'contingency table'. The above example is a 2 X 2 crosstabulation (which therefore has four cells). A similar exercise could be performed if *class* were coded into three values (say, 'middle class', 'non-manual working class' and 'manual working class'). The resulting crosstabulation would be a 3 X 2 table containing six cells.

A variety of statistical tests can be performed with such a two-variable classification to establish whether there is a significant statistical relationship between the two variables concerned (here *sex* and *class*).

SPSS/PC+ will crosstabulate such information, and analyze it statistically, using the command **CROSSTABS**.

For the above example the instruction would be:

```
CROSSTABS  SEX BY CLASS.
```

This would produce a crosstabulation table, labelled with any variable labels and value labels which had previously been provided. The number of cases with relevant values for both of the variables would be displayed in each cell. Thus the final count of all cases in the sample which had been coded as 'male' on the *sex* variable and as 'working class' for the *class* variable would be displayed in the 'male working class' cell.

Associated with many SPSS/PC+ commands are a number of **sub-commands**. These have various uses but often have the function of allowing the user to tailor the form which an analysis will take, or the form in which output will be displayed.

Thus for **CROSSTABS** there is an optional subcommand **Cells** which can be used to specify additional numerical information to be included within the output display. Thus the subcommand 'Cells = count row' will produce an output table in which the count within each cell *and* the row percentages will be displayed (without such a command only the *default* display would be produced, with counts for cells, rows and columns, but no percentage calculation). Another optional subcommand, **Format**, specifies how the output table is to be drawn and labelled (thus "Format = nolabels" will suppress both value and variable labels). The **Statistics** subcommand can be used to request that statistical tests be performed on the data and the resulting values displayed on the output table. For example "Statistics = chisq" will cause a statistical test known as the chi-square test to be performed on the data included in the crosstabulation. The subcommand "Statistics = chisq uc" will lead to both a chi-square value and an 'uncertainty coefficient' value being added to the table. Thus if we wished to use **CROSSTABS** to print a table which includes counts and row percentages, which excludes variable- and value-labels, and which includes both the chi-square value and the uncertainty coefficient, the appropriate command would be:

```
CROSSTABS  SEX BY CLASS / CELLS = COUNT ROW / FORMAT = NOLABELS
                          / STATISTICS= CHISQ UC.
```

(Notice that the subcommand is preceded by a slash /. This is an example of a rule of SPSS/PC+ syntax. This will be fully explained later – remember that the function of this chapter is merely to provide an overview).

It can be seen that by using appropriate subcommands the user is able to control many aspects of the procedure. The subcommands available, and the options available for any one subcommand, vary considerably depending on which command is being issued.

Sometimes we would wish to split our population of respondents into two groups (for example, 'males' and 'females') and then test to see whether there is a difference between the groups on a variable such as *age*. The appropriate statistical test for examining such a difference is the *t*-test, and the appropriate SPSS/PC+ command to request such a test is **T-TEST**. To find out whether there is a significant age-difference between the males and the females in our sample we would need to use the variables *age* and *sex*.

Within the command to be issued we will need to specify the two groups – males (coded 1 on *sex*) and females (coded 2), using:

```
GROUPS = SEX(1,2)
```

and we will also need to specify the variable (or set of variables) on which we wish to compare our groups, using, for example:

```
VARIABLES = AGE
```

Incorporating these elements into a **T-TEST** command, using the appropriate SPSS/PC+ syntax, yields the command:

```
T-TEST   GROUPS=SEX(1,2) / VARIABLES= AGE.
```

If we wished to perform *t*-tests to compare males and females on age, length of education (say, *leduc*) and annual salary (say, *income*), this could be specified in a single command:

```
T-TEST   GROUPS=SEX(1,2)/VARIABLES= AGE LEDUC INCOME.
```

For those who are statistically sophisticated, let me add, to whet your appetite, that SPSS/PC+ provides for a whole range of 'non-parametric' analyses, anova, correlations, and regression analyses, etc. It will also plot data graphically in a variety of ways.

For the *really* sophisticated I will mention, too, that other modules in the SPSS/PC+ system will also perform – with equivalent ease of operation – factor analyses, discriminant and cluster analyses, and log-linear modelling.

And if such terms mean little to you at this time, relax! SPSS/PC+ will allow you to explore your data to the full, and with instant feedback you will be able to fully examine the characteristics of your dataset as few were able to do before the advent of microcomputer based versions of the SPSS system. Your statistical literacy and your statistical skills will increase, and the depth and quality of your analyses and of your reports will improve significantly.

Remember that the purpose of this initial overview is to present the general framework for using SPSS/PC+. Do not try to learn the details at this stage – and do not worry if the statistical terms (like '*t*-test' and 'anova') mean little to you. The aim at this stage is to give a general idea of how the system can be used. Those who have some familiarity with statistics, however, may gain an appreciation of how easy it is, once a dataset has been established, to perform otherwise complex and cumbersome analyses.

By now you should have some understanding of the stages involved in creating a dataset for use by SPSS/PC+. It is *nct* important that you *remember* the details contained in this chapter, but it *is* important that you have *understood* all of the material. In the following chapters we will be building on this basic understanding.

You should now have a preliminary understanding of how:

- raw data are collected and coded into numerical form
- numerical data are entered into the computer and saved
- these data are annotated with descriptions and labels
- the annotated data are used to produce summaries and tables, and to perform statistical tests

If you feel unsure about any of the general concepts it would be well to re-read one or more of the sections in this chapter before going on to the next. Remember, too, that consulting the comprehensive glossary in *Appendix G* will help you to understand any terms and concepts that are unfamiliar to you.

When you are confident that you have a general grasp of the SPSS/PC+ system, as outlined in this chapter, it is time to proceed to Chapter 2.

2 *Collecting and Coding Data*

This chapter deals with the first of the four stages outlined in Chapter 1. In this chapter we learn how to collect data and code them into numerical form.

The discussion will include a consideration of:

- variables, cases and values
- coding strategies
- preparation of the questionnaire and 'codebook'
- coding the raw data

2.1 Variables, cases and values

Typically, SPSS/PC+ will be used to analyze data obtained from a survey in which we have collected many items of information (**variables**) from (or about) a number of different people (called **subjects** or **respondents** or, more generally, **cases**). Alternatively, our dataset might consist of data relating to variables associated not with individual people but with cases of other kinds – different products, for example, or different countries. The aim, in collecting and coding data, is to arrive at a situation in which we have, for each **case**, a **value** for that case on each **variable** in which we are interested.

Thus for a **variable** which has been given the name *age* we might have the **value** 35 for **case** 1 (John Doe) and the value 23 for case 2 (Sarah Smith). For the variable *income* we might have the value 12 000 for case 1 (John Doe) and the value 10 000 for case 2 (Sarah Smith).

The data to be analyzed are organized into a grid in which each case occupies a separate **row**, and each variable occupies a vertical slice of the appropriate width (see Table 2.1).

Table 2.1

Case	**Variable 1 Age**	**Variable 2 Income**	**Variable 3 Height**	**Variable 4 Weight**
001	35	1750	72	61
002	23	1200	68	41
003	25	2100	70	54
004	27	2335	63	38
etc.				

Here each **case** is represented by a row of data. Each **variable** is represented by a vertical 'slice' of the data matrix which is either two single columns in width or (for variable 2) four single columns in width. For each **case**, a **value** is associated with each **variable**, and these values (16 in the example provided) constitute the **data** for analysis. We have data on each of four variables for each of our four cases, making 16 data points within our dataset.

A typical SPSS/PC+ project may involve hundreds of cases, and up to 500 variables may be represented within a single dataset.

2.2 Coding strategies

Most of the examples provided in this book use data from a survey in which a number of students (cases) responded to a written questionnaire. The Gender Attitude Survey is briefly described in *Appendix C*. To analyze the results of this survey each questionnaire *item* is treated as a separate **variable**. Data are available on all of the variables for each **case** (each of the 50 students who took part in the survey), and each person's response to each item will need to be coded into an appropriate **value**.

Some of the data will not need to be subjected to any special coding procedure because they will actually be **collected** in numerical form (e.g. age in years, or the number of children in a family). Other raw data will need to be transformed into an appropriate coded value. Throughout most of this book it will be assumed that data are coded in terms of whole numbers ('integers'). Although SPSS/PC+ can handle alphabetic and non-integer data (i.e. data containing decimal points) the use of such data will not be considered until Chapter 16.

Data which are *not* collected in numerical form will therefore be coded into whole numbers. Thus sex (**male** or **female**) may be coded numerically so that one number (say, 1) signifies 'male' and another (say, 2) signifies 'female'. Such numerical coding can be achieved for almost all data. Thus for a question to which respondents are invited to answer 'Yes', 'Not sure' and

'No', we could code the answers as 1, 2 and 3 respectively.

The actual coding system used to translate answers into numerical values may be arbitrary, but we must keep a clear record of our 'key' for translating the responses into numbers. And in coding any particular variable (such as the answers to one particular questionnaire item) we must apply precisely the same coding system to each case (or respondent).

Thus, in coding the sex of a respondent into a numerical value, instead of 1 = 'male', 2 = 'female' we could equally well use 1 = 'female' and 2 = 'male'. But, once the rule has been established, the sex of *all* subjects must be coded according to the *same* rule.

Another example of numerical coding is provided by the case where we ask respondents how much they agree or disagree with a statement using a fixed range of levels of agreement. Thus:

'Do you agree that women should have the same rights as men?'

Respondents could be asked to indicate how much they agree or disagree with this statement by ticking one of five responses:

... Strongly agree
... Agree
... Neither agree nor disagree
... Disagree
... Strongly disagree

For analysis, the answer each respondent gives to this question could be coded numerically (for example, using 1 to indicate 'Strongly agree', 2 to indicate 'Agree', 3 to indicate 'Neither agree nor disagree', 4 to indicate 'Disagree' and 5 to indicate 'Strongly disagree').

So far, each of the examples has involved a closed-ended (forced-choice) question. Another example would be "Would you prefer (a) a restful holiday, (b) an active holiday or (c) something in between"? We could code the responses (a), (b) and (c) as 1, 2, and 3 respectively.

Sometimes, however, we might ask a more open question. Thus:

'What is your favourite hobby? (please state) ....'.

Here we are inviting (and will get) a huge variety of answers, from alchemy to zoo-visits. For any tabulation or statistical analysis, we have to categorize these answers. The best way, for our purposes, is to establish particular categories and to assign a different code to each. This can be done in a number of ways. Let's take two extreme examples:

(a) We could simply decide to code any hobby as either 'outdoor' (code 1) or 'indoor' (code 2).

or

(b) We might decide to wait until all questionnaires had been returned and then list all the answers obtained. We could then provide a coding number for each hobby mentioned as the favourite hobby by at least one person. Thus if 100 respondents produced a collection of 57 different 'favourite hobbies', each hobby could be given a code number (01 to 57).

Between these two extremes we might decide (either before conducting the survey or when we have had a chance to look at the range of answers) a limited number of categories for coding. Thus we might divide all possible favourite hobbies into the following categories:

1. Sports/games
2. Social
3. Collecting
4. Crafts
5. Music
6. Other.

We could then code the data using the numbers 1 to 6. Sometimes a difficult choice would have to be made. For example, should 'collecting and listening to Beatles' albums' be coded as 3 or 5? Because only one value (or code) is permitted for each respondent for each variable it would *not* be possible to use both. Notice also the 'other' category, used for the eccentric cases which don't fall easily into one of the set categories. Some such 'miscellaneous' category is usually needed when we attempt to code responses to open-ended questions.

There is another issue which needs to be raised. It is inevitable that some of the respondents who return a questionnaire will fail to return a complete set of responses. A subject may make the error of missing out an item or may turn two pages at once. Subjects sometimes object to a question and refuse to answer, or give an uncodable (and sometimes unprintable) response rather than choosing from the range of responses provided. In all such instances the data are treated as 'missing'. The way to deal with what would otherwise be a 'hole' in the dataset is to provide an additional code which is used only to specify that the relevant data point is missing from the dataset. Any missing data are then coded with this **missing value**. For example, if *sex* is coded as 1 for 'male' and 2 for 'female', the value 0 (or perhaps 3, or 9) can be assigned as the missing value.

As part of the process of defining the dataset, the system can then be instructed that a particular value for a variable represents not real data but missing data. The SPSS/PC+ system recognizes such an instruction and can deal appropriately with missing data (for example, if it has been instructed to treat '99' as the missing value for the variable *age* it will *not*, in calculating the average age of respondents, include any '99' values as if there were people in the sample who were just one year short of a century!)

Finally, note that some open-ended questions will be so varied that any simple coding organization will be difficult to devise. For example:

'What do you remember of your first day at school'?..............

..

The data which will emerge from such an open-ended question will often be rich and interesting, but they will not be in an ideal form for statistical analysis.

Contrast this open-ended question with the following:

'How do you recall your first day at school? (tick ONE)'

... Interesting
... Boring
... Neither boring nor interesting
... Cannot remember

This is easily coded, and will therefore be amenable to statistical analysis (although it has to be admitted that the data which will emerge will not be nearly as full or interesting, in this case, as those which would be likely to emerge from the open-ended question).

The raw data from responses to written questionnaires, interviews, etc. are therefore prepared for use by SPSS/PC+ by being coded according to a uniform coding scheme. All of the information is thus reduced to a series of numbers.

Some of these numbers will be real numbers (like age in years) and represent **numerical variables**. Other numbers, however, will be code numbers (like 1 = 'yes, 2 = 'no', 9 = 'missing'). These represent the data from **categorical variables**. Codes are assigned to the various categories by the user and raw data relating to the particular variable are then categorized and coded appropriately.

The user provides the computer with the dataset in the form of a matrix of numbers (where the columns are organized into single-column, two-column or larger vertical slices, each representing a single variable, and where each row – or for extensive datasets each group of two or three rows – corresponds to a particular case). The variables may be derived from questionnaire items of varying style and format, but the information relating to each variable is coded into numerical form and placed in the appropriate column(s) of the dataset. Thus, for a sample of cases, each variable corresponds to a vertical slice through the dataset (or data matrix).

2.3 Preparation of the questionnaire and codebook

When developing a questionnaire it is useful to write out a **codebook** – a list of the variables corresponding to the items of the questionnaire, together with information about how data relating to each of the items will eventually be coded. Decisions may be made at this stage about such matters as how many digits (columns) will be needed to code a particular variable, and what codes will be assigned to particular answers. Such information may enable the person designing the questionnaire to include information, on the questionnaire form itself, that will later make the task of entering data into the dataset swift and accurate. Thus some questionnaires include a separate column (which the respondent is asked, e.g. to **PLEASE LEAVE BLANK**), containing aids to scoring the questionnaire. Such an arrangement enables responses to the completed questionnaire to be coded directly on the form and later copied onto special **data sheets** or entered directly into a computer **data file**.

Preparing a questionnaire

The first step is to construct an initial draft of the questionnaire. It is useful, even at this stage, to think in terms of the variables which will be produced. Certain types of analysis demand **numerical** rather than **categorical** variables, for example, so the form a question takes may be partly determined by the type of analysis needed. The questionnaire should be comprehensive enough to cover the full range of variables required, without being over-long. It should have a definite logical structure, and the order of questions should be carefully considered. 'Branching' may be included (e.g. **'If YES then go straight to Question 7 – otherwise answer Questions 5 and 6'**), but such complexities should be kept to a minimum in questionnaires which are to be completed in writing by the respondents (and should be avoided altogether if at all possible).

The user needs to decide whether questions are to be open-ended or closed, and any closed questions demand careful consideration regarding the range of alternatives to be provided. If only one answer is to be permitted to a question, the alternatives presented should be **mutually exclusive** (i.e. only one of them *can* apply in any one case). The alternatives should also be **mutually exhaustive** (i.e. one or other of them *must* apply to each case).

An example of a question which breaks both of these rules is:

'What is your current marital status?':
Married / Divorced / Previously married / Widowed

This question does not provide mutually exclusive alternatives because a divorced person could validly tick both 'divorced' and 'previously married'. Also, for a general population, the alternative answers would not be mutually exhaustive, because no provision is made for a response from a single (= 'never married') person.

Another golden rule in questionnaire design is to avoid ambiguous phraseology. The question **'Were you married last year?'** for example, could mean **'Did you *get* married last year?'** or **'Were you of 'married' status last year?'**. Thus a person who had been married for the past ten years could equally well answer 'no', in response to the first meaning, or 'yes', in response to the second meaning. The question is also ambiguous in another way. The term 'last year' sometimes means 'the past 12 months' and sometimes means 'during the last calendar year'.

Some of the more subtle ambiguities (including words or phrases which have different shades of meaning for different people) can only be identified as the result of piloting an early draft of the questionnaire with respondents similar in age, social class, and educational level to those who will form the real survey population. The research worker needs to talk at some length with people who have attempted to complete the questionnaire, to discover any problems they may have had in understanding or answering the questions.

Before the final version of the questionnaire is typed or printed, and multiple copies made, it is important to consider whether provision is to be made for coding the data on the questionnaire itself. If 'on-form' coding is to be performed the inclusion of information from the codebook might aid the task of the coder. Such information, and any spaces provided for coding, should be clearly separated from the part of the form which the respondent is expected to complete. A line should be used to separate the two 'fields' (the respondent field and the coder field) and the coder field should be headed with a message such as **'PLEASE LEAVE BLANK'**.

The Gender Attitudes Questionnaire (*Appendix C*) provides an example of how a questionnaire may be prepared to facilitate easy on-form coding. Such aids to coding, however, can be included only after the codebook has been compiled.

Preparing a codebook

When the questionnaire itself has been prepared in its final draft, it needs to be examined carefully and a codebook prepared. The codebook is essentially a list of variables with information about each. It is a written guide to the variables to be extracted from the questionnaire and is for the information of the user and other people who might be involved in coding variables or entering data into a computer file. It is *not* part of the SPSS system and is not a computer file (although it may be produced with the aid of a word-processor). It is 'for human consumption only'. An example of a codebook (produced for the Gender Attitudes Survey) is provided in *Appendix D*.

The codebook will include a list of all of the variable **names**, the valid values of data for each variable, and a value to be assigned to data which are missing. Useful additions include a note of the nature of the data (e.g. **C** for categorical variables such as *class* or *sex*, and **N** for true numeric variables such as *age* in years or *height* in inches). It is also useful to include a brief description of the variable and a key to the coding scheme to be used in coding each variable. This information is likely to correspond to any **variable labels** (remember that these are *not* the same as variable **names**) and **value labels** (e.g. 1 = 'male', 2 = 'female') that the user intends to enter at a later stage into an SPSS/PC+ file known as the **data definition file**.

The questionnaire first needs to be **itemized**, each item yielding a single **variable**. In most cases this does not present a problem. Many questions clearly produce a single variable. Thus:

(a) 'Women should have the same rights as men'
Strongly agree / Agree / Neither / Disagree / Strongly Disagree

(b) 'Current occupation' ....
(to be coded in terms of one of five social class categories)

(c) 'Which of the following is your *favourite* colour?'
Red / Orange / Yellow / Green / Blue / Indigo / Violet

(d) 'For how many hours in an average week do you watch television?' ......

(e) 'Which ONE of the following games would you prefer to engage in':

Hockey ...	**Football ...**	**Skiing ...**	**Squash ...**
Tennis ...	**Shooting ...**	**Golf ...**	**Chess ...**

In each of these examples only *one* answer is permitted – and thus each of the five questions (a) to (e) provides a single variable. Consider, however, a variation of example **(e)**:

(f) 'Which of the following games have you engaged in during the past FOUR WEEKS':

Hockey ...	**Football ...**	**Skiing ...**	**Squash ...**
Tennis ...	**Shooting ...**	**Golf ...**	**Chess ...**

Here some respondents may tick *none* of the games listed. Others may tick *one*, and some will tick *more than one*. We could code the answers to this question in terms of a single variable by merely counting the number of ticks given – 0 to 8 – but this form of coding would clearly not capture the full details of the data available on a completed questionnaire. We might wish, at some future stage, to count the number of respondents who had engaged in *each* of the individual games. To preserve the full information from respondents' answers we should treat question **(f)** as containing eight separate items, and hence code the answers in terms of eight separate variables.

Thus the variable *hockey* might be coded 0 if there was no tick present, and 1 if a tick had been given (and so for the other seven games). If we later wished to count the total number of listed games engaged in by a particular respondent during the past four weeks, we could use a special **COMPUTE** feature of the SPSS/PC+ system to add the number of ticks given by that respondent in response to question **(f)** (i.e. the number of variables, among the eight, which were coded with a 1 – meaning 'game engaged in'). If we had opted to include all of the information from the question in terms of a single variable, however, and merely counted the ticks, there is no way in which we could later unscramble the resulting number (0 to 8) to determine how many respondents had engaged in, say, hockey. It is worth bearing in mind the old saying 'You can't unscramble eggs'. The best policy is to include the original data in as full a form as you will possibly require, and this might mean that a single question will yield not one but several variables.

When we have itemized the questionnaire we will have produced a list of variables. We should then consider, for each variable in turn:

(i) a suitable name for the variable, to be used by SPSS/PC+
(ii) the variable label to be assigned (if any)
(iii) the variable 'type' (numerical or categorical)
(iv) coding values to be given to the variable
(v) the range of valid values for the variable
(vi) the value labels to be assigned (if any)
(vii) the value to be assigned to any data which are missing
(viii) the number of columns which the variable will occupy within the data matrix
(ix) the position of these columns

Remember that we are still working with pen and paper to write our codebook, although the decisions made at this stage will determine how we code our data, how we arrange the data to form a dataset, and how we define and label our dataset when we come to write our SPSS/PC+ **data definition file**. Thus compiling the codebook is a very important step undertaken *before* we approach the computer.

(i) **Variable name:** Valid variable names in SPSS/PC+ consist of up to eight characters (letters or numbers), the first of which must be a letter. Names are often chosen to be descriptive of the variable (like *sex* and *age*) although it is often simpler to opt for a less specific formulation (thus *ques1*, *ques2*, *ques3*, etc., or *var1*, *var2*, *var3*, etc.). Several words have a special function in the SPSS system (they are **keywords**) and should not be used as variable names. The keywords which are reserved for use by SPSS/PC+ are ALL, AND, BY, EQ, GE, GT, LE, LT, NE, NOT, OR, TO and WITH.

(ii) **Variable labels**: It is not always easy for the user to identify a variable from the variable name (particularly if a name such as *var001*, *item10*, or *ques5* has been assigned). The codebook should therefore include a brief description of each variable (thus *ques5* might be given the description 'Anxiety experienced during exams.'). SPSS/PC+ allows the user the option of providing a **variable label** for some or all of the variables, and it may be useful to devise a codebook description which could also be used as the SPSS/PC+ variable label. When such labels are entered into the SPSS/PC+ **data definition file** the system uses them to annotate the output from various procedures, and this can make it considerably easier to understand the output. Variable labels should be devised to provide maximum information to the user, but should not be too long. Some thought should therefore be given to such labels as the codebook is being compiled. SPSS/PC+ variable labels may be up to 40 characters in length and may include spaces. Remember, though, that labels do not *have* to be supplied to the SPSS/PC+ system. Variable labels are optional and their presence or absence has no effect on the analyses performed.

(iii) **Variable type**: Different kinds of analysis are appropriate to categorical data (such as that relating to country of origin, sex, or favourite sport) and numerical data (such as that relating to age in years, annual income, or number of children). It may prove useful to include within the codebook a reminder of which variables are categorical and which are numerical. Some variables (such as *sex*) are clearly categorical and others (such as *age* in years) are clearly numerical. However, *age*, instead of being represented numerically ('age in years' – 01 to 99) could be classified so that it becomes a categorical variable (e.g. 1 = 'young' = 30 or younger; 2 = 'middle' = 31 to 50; and 3 = 'old' = 51 or over).

Adding a **C** or **N** in the codebook for each variable may act as a useful reminder of the variable type at a later stage when complex analyses are being planned.

It should be noted that it is not always easy to decide whether a variable should be considered as categorical or numerical. For example, some variables may be represented by a number corresponding to the number that a respondent has circled or ticked on a 5-point or 7-point scale (the ends of such a scale might be labelled 'totally agree' and 'totally disagree', or 'extremely anxious' and 'not at all anxious'). The issue is whether such a response should be treated as a measurement (i.e. a numerical value) or merely as a category (i.e. a coded value) is the subject of considerable debate. (In statistical analysis, different criteria apply to true numbers and to codes. It is permissible to calculate the arithmetic average for true numbers, for example, but codes should not be averaged.)

Some people would argue that the fact that numbers are used in 5-point and 7-point scales does not make the variable numeric, and that responses to such scales should be treated only as codes. Others, however, are happy to treat such data as numeric. Information concerning the variable type is for the user's reference only. It may be entered in the codebook but is not at any stage entered into the SPSS/PC+ system. Thus any variables which are ambiguous with regard to their categorical/numeric type may be signalled in the codebook with a question mark – **?**. Decisions regarding which analyses are appropriate to that variable may need to be considered carefully at a later stage, perhaps after relevant features of the variable (such as its **frequency distribution**) have been examined in preliminary SPSS/PC+ analyses.

(iv) **Coding values**: Data in categorical form will need to be coded. For example, the user needs to decide for the variable *sex* what code is to be used to denote 'male' and what code is to be used to denote 'female'. Typically the code 1 would be assigned to one sex and the code 2 to the other. Where a list of possible responses has been provided for the respondent, the codings would typically start at 1 for the first possible response and continue through 2, 3, 4, etc. until all possible valid responses have been assigned a different number. Thus where the respondent has been presented with a statement and asked to indicate in terms of a fixed range of possibilities how far he or she agrees or disagrees with the statement, 1 might be assigned to 'Strongly agree', 2 to 'Agree', 3 to 'Undecided', 4 to 'Disagree' and 5 to 'Strongly disagree'.

Devising a coding scheme for closed-ended ('fixed-choice') questions usually presents no problem but devising a coding scheme for more open questions is more demanding. If a variable such as *currocc* (= current occupation) is to be coded then a **coding frame** needs to be established. Thus occupation might be classified in terms of social class, or according to some other categorization system (e.g. 'indoor work', 'outdoor work', 'both indoor and outdoor'), or a large range of individual occupations might be listed as categories. Whichever categorization system is adopted, a comprehensive list of categories must be devised, and a unique coding number assigned to each category. Coding systems should be both **exclusive** (no categories overlap) and **exhaustive** (all valid responses fall into one or other of the categories provided). A general category (such as 'Other') might be provided to ensure that the category system is exhaustive.

Standard coding schemes are available for some variables (including the classificatory system used to code occupation in terms of social class, and certain systems used to classify racial origin) but for many variables the researcher may need to establish an original coding scheme. The quality of a coding scheme depends to a large degree on the skill of the researcher, and his or her experience with the topic under investigation. In some cases it is advisable to devise the coding scheme only after the questionnaires have been collected and examined, so that the researcher has some idea of the range of responses obtained.

When the comprehensive list of categories relating to a particular variable has been constructed it only remains for each category to be assigned a unique code number. Each of the categories, together with the relevant code number, should be included within the codebook.

(v) **Range of valid values**: It is also useful to include a note of the **valid range** of values for each variable within the codebook. If preliminary SPSS/PC+ analyses of a dataset then reveal the presence of values outside the valid range (and not corresponding to an extra value assigned to missing data) this will indicate that coding or typing errors have been made.

(vi) **Value labels**: When drawing up the codebook listing of possible categories for any variable, together with the relevant code numbers, it is useful to bear in mind any **value labels** that you may wish to include in the SPSS/PC+ **data definition file**. Like variable labels, these are entirely optional and have no effect on SPSS/PC+ analyses, but SPSS/PC+ output is more easily intelligible if labels are provided for some or all of the values which the variable may take. Thus for the variable *sex*, we might provide the label 'male' for the coding 1 and the label 'female' for the coding 2. This would mean that instead of a table indicating, for example, the average income for '1' and the average income for '2', the table would present an average income for 'male' and an average income for 'female'. With a more extensive category system the advantage of using labels would be even more pronounced.

Thus appropriate value labels often make the task of interpreting tables and the results of analysis simpler, and it is not too soon to be thinking about such labelling when you are compiling a codebook for your project. SPSS/PC+ permits value labels up to 60 characters in length, and spaces may be included, but it is best if such labels are kept reasonably short. In some cases only *some* values for a variable will merit labelling. Thus it may be sensible to label only the extreme values obtained by using the traditional 7-point scale (e.g. 1 = 'strong agreement', 7 = 'strong disagreement').

(vii) **'Missing' values** : A special value should be assigned to each variable for coding missing data. In preparing the codebook you should assign one particular value (for example the number 9 or 0, or perhaps 99 for a two-digit variable) as the 'missing' value. The value assigned as the missing value should of course be one which is not otherwise used to code data for this variable and which is therefore *outside* the valid range. If, during coding, missing or 'spoiled' data are encountered, the missing value(s) should be entered into the columns assigned to the relevant variables. Coding with valid values will resume when legitimate responses are once again present on the questionnaire.

(viii) **Data columns**: When the questionnaires have been completed by respondents in the study, the value for each variable will need to be represented within the dataset. Sometimes an actual numerical value will be included (as for *age* – 'age in years' or *nochil* – 'number of children'), but the values included for many variables are likely to be codes (thus, for the variable, *sex*, 1 might be entered into the dataset for each male subject, and 2 might be entered for each female subject). The **data columns** item of the codebook notes how many columns within the dataset (or data matrix) need to be assigned to each variable.

Variables which have 10 or fewer possible values (*including* the 'missing value') can be reported or coded in terms of a single digit (0–9) and can therefore be represented within a single column within the data matrix. Variables which have more than 10 possible values, but fewer than 100 (again, *including* the 'missing value'), can be accommodated within two columns (i.e. 00, 01, 02, etc. to 99). A variable such as annual income (reported in terms of pounds sterling, or dollars, for example) may need five or six columns assigned to it.

Each variable, therefore, should be assigned the appropriate number of single-digit columns within the data matrix. This might be reported in the codebook as '*sex*–1 column', '*age*–2 columns' (unless we feel that we have to allow for the possibility of respondents over 99-years old, in which case 3 columns should be assigned), '*income*–4 columns', etc. Remember in each case to allow for a value to represent any missing data, so that a variable which can have 10 *valid* values will have 11 possible values altogether and would therefore need to be assigned *two* columns within the data matrix.

(ix) **Column position**: Once we know the number of columns needed to accommodate the variables in our list we can begin to assign *particular* columns within the data matrix to particular variables. Thus if the first variable – *id* ('identification number') – is assigned three columns, these can be columns 1 to 3 (inclusive). The next variable (say, *sex*, accommodated in just one column) can then be represented in column 4. We can therefore map out the full specification of the data matrix by following the sequence of variables precisely, and assigning at each stage the next column for a single-column variable, the next two columns for a two-column variable, and so on. The position of the column(s) occupied by the variable should be entered in the codebook. Thus:

id	Columns 1–3
sex	4
age	5–6
income	7–11

Notice that the column numbers specified are **inclusive**. Thus the specification '7–11' represents **5** columns (7, 8, 9, 10 and 11).

Instead of assigning the immediate next column(s) to the next variable, as in the above example, it is possible to leave one or more blank spaces between each variable. This separates the data for different variables very clearly in the data matrix and can therefore make a printed copy of the data matrix easier to read. However, with a large number of variables relating to each case, the inclusion of separation spaces means that the line of data relating to a case will soon be filled, and that two or more lines will sometimes be needed to accommodate data that could otherwise be accommodated on a single line. The inclusion of separation spaces also adds to the tedium of typing in the dataset. It's your choice. The point is that spaces *between* data items are permitted.

SPSS/PC+ works with an 80-column format which has become the standard for much computer work. Thus each 'line' (or row) within the data matrix can be no more than 80 columns in length, although data from a single respondent may occupy two, three or more successive rows (referred to as 'records') within the data matrix. It is important to bear the 80-column limit in mind when assigning columns to variables. If you are approaching column **80** and have a variable which spans two or more columns you should stop at a suitable column number (say **77** or **79**), to avoid data for a single variable beginning on one line and ending on the next (such an arrangement is *not* permissible within SPSS/PC+).

If you still have variables to assign to columns when the first data line is full (or as full of complete variables as it can be) you should now begin assigning columns on the second line. Column numbering begins again at column **1** (*not* **81**), so make a note (within the codebook) that column numbers now represent a second record (line of data) *for the same case.* Sometimes three, four or even more records may be needed to accommodate the data for a single case. When assigning columns to data lines other than the first, it is good practice to 'indent' the line slightly (i.e. assign the variable that starts the second, and any subsequent, line to begin not in column 1 but in, say, column 4). This makes it easy to locate particular cases on the printed record of the data matrix. *All* cases (respondents), of course, will be represented by the same number of records, and the data relating to each respondent will end in precisely the same column of the same record. The codebook specifications form a basic template which applies to the data for all respondents.

Using the examples (a) to (d), and (f)(p. 20) as if they appeared in that order at the beginning of a single questionnaire (preceded only by the variable *id* – 'subject identification number') the Table 2.2 illustrates one form in which the corresponding codebook might be produced.

Table 2.2

Variable name	Col/s	'Type'	Values	Value Labels	Miss.	Var. label
ID	1–3	N	–	–	999	Subject No.
EQRIT	4	C	1	strongly agree	9	Agree with equal rights
			to			
			7	strongly disagree		
OCCUP	5	C	1	manual	9	Occupation – social class
			2	clerical		
			3	manager		
			4	profess.		
FAVCOL	6	C	1	red	9	Favourite colour
			2	orange		
			3	yellow		
			4	green		
			5	blue		
			6	indigo		
			7	violet		
TVHOURS	7–8	N	–		99	Number hours of T.V./week
FAVGAME	9	C	1	hockey	9	Preferred sport from 8 supplied
			2	football		
			3	skiing		
			... 8	chess		
HOCKEY	10	C	1	played	9	Hockey in past month?
			2	not played		
FOOTBALL	11	C	1	played	9	Football in past month?
			2	not played		
SKIING	12	C	1	played	9	Skiing in past month?
			2	not played		
etc.						

2.4 Conducting the survey

The next step is to produce multiple copies of the questionnaire and to distribute these to a suitable respondent population. Special care is needed with sampling, particularly if the survey is aimed at reaching a representative sample of a wider population (such as 'unmarried males' or 'working class wives'). Complex strategies are available for ensuring that samples are representative and if your survey is of this type then you would do well to consult a specialist text. Initially, you might consult the **sampling procedures** entry in the Glossary (*Appendix G*).

If the questionnaire is not to be completed by a captive population such as a class of students it may be necessary to arrange for pre-addressed envelopes to be included with the questionnaire, or pick-up points to be specified, in order to encourage a high rate of return. Unless the population is captive a 100 per cent return rate cannot be expected, and it is possible that the respondents who fail to return their questionnaires will be different in some important way from those who did, thus limiting the extent to which we can generalize from the analyses obtained.

2.5 Coding the data

When the completed questionnaires arrive you can begin to code the data using the codebook. The aim is to reduce all of the information in each questionnaire to a single string of digits for each case. Sometimes it will be possible to enter the information directly into a computer **data file** from the questionnaire. With more complex questionnaires, however, several stages may be necessary.

For example, a respondent's answers may first be coded onto the **coding field** of the questionnaire itself (e.g. into the extreme right-hand part of the form, perhaps labelled 'Please leave blank'), and later transferred to special data sheets, to be entered at a later stage into a computer **data file**. Data sheets are simply empty grids, typically of 80 columns by 25 rows, with the columns numbered at the top of the sheet and with useful dividing lines marked after every 10th column. Using such a grid helps the coder to make sure that data relating to each variable are placed in the appropriate column(s). Because each subject has been given the same questionnaire, a parallel set of data must be obtained for each case. Thus some value must be included (even if it is a missing value) for each respondent on each variable. Thus the data matrix obtained as a result of the scoring process will take the following rectangular form:

```
0011191432421342132341411412412342142314234
0022223213131321323312313123123131313131222
0031234234333343424333333334225653455555444
0042193434444445454444453211244314232332344
0052202342342342423423423444562342344444233
0061234323423423432423424242424242342342344
```

In this case, each row across the page represents the data relating to a single respondent. When all of the data relating to a single case cannot be accommodated on a single row (record) two or more rows will be necessary to record the data for each subject.

If the data for a single subject occupy two records, the data matrix may appear in the following form:

```
0011191432421342132341411412412342142314234676576571623561253331236
   1212221111111222222211222
0022223213131321323312313123123131313131222334612543654352433462546
   2221212222211122121122222
0031234234333343424333333334225653455555444342165436541654654654615
   1212121212222212222122222
0042193434444445454444453211244314232332344425425425435412354215421
   2212222111221212121222221
0052202342342342423423423444562342344444233132432432121255565465465
   2122212222212111122222111
0061234323423423432423424242424242342342344126565465525242654254652
   1222212121212222222211111
```

It can be seen that, although the data from each subject now occupy two lines, the rectangularity of the grid is maintained (note also how the second record for each case has been indented).

Returning to the example of the shorter matrix, we can see that the first respondent in the study (given the identification number '001') has supplied a set of answers which have been coded into the following row of numbers:

```
Subject 1        0011191432421342132341411412412342142314234
```

The data from subjects 2 and 3, who produced somewhat different answers to the same set of questions, are given in the second and third rows:

```
Subject 2        0022223213131321323312313123123131313131222
Subject 3        0031234234333343424333333334225653455555444
```

A single **column** (or, sometimes, two or three columns together) will always include the coded values for the answers given to a particular questionnaire item (or other variable).

Notice that the matrix is perfectly aligned on the right edge. A 'ragged right edge' to the matrix, like this:

```
Row 1            0011191432421342132341411412412342142314234
Row 2            0022223213131321323312313123123131313131222
Row 3            00312344234333343424333333334225653455555444
Row 4            00421934444445454444453211244314232332344
Row 5            005220232342342423423423444562342344444233
Row 6            0061234323423423432423424242424242342342344
```

would indicate that mistakes have been made in entering the data. Perfect alignment of columns throughout the data matrix is essential because the computer will take all numbers in any column (or, sometimes, in a group of two or three columns together) as representing the same variable (e.g. answers to the same question) for all subjects.

Using our sample matrix again, we can number the columns 1–43.

```
     COLUMN                  1111111111222222222233333333334444
        NUMBERS 1234567890123456789012345678901234567890123

 (Subject 1)    0011191432421342132341411412412342142314234
 (Subject 2)    0022223213131321323312313123123131313131222
 (Subject 3)    0031234234333343424333333334225653455555444
 (Subject 4)    0042193434444454544444532112443142323332344
 (Subject 5)    0052202342342342423423423444562342344444233
 (Subject 6)    0061234323423423432423424242424242342342344
```

Now let's examine in some detail just the first 10 columns of this dataset. For the sake of clarity, the following display includes spaces (these do *not* correspond to separation spaces within the actual data matrix) placed between the each pair of slices in the dataset. Each slice corresponds to the data for a single variable:

```
 COLUMN                                1
 NUMBERS        123  4  56  7  8  9  0

 (Subject 1)    001  1  19  2  4  3  2
 (Subject 2)    002  2  22  1  2  1  3
 (Subject 3)    003  1  23  4  2  3  4
 (Subject 4)    004  2  19  1  4  3  4
 (Subject 5)    005  2  20  1  3  4  2
 (Subject 6)    006  1  23  4  3  2  3
```

In this example the first three columns (**1, 2 and 3**) contain the respondent's identification number (the variable *id*). The use of three columns means that we can accommodate up to 999 subjects with separate *ids*. If we had a smaller respondent group (less than 100) we could make do with two columns (*id* values from 01 to 99), but since we may want to extend our survey later it might be best to keep the *id* format to permit up to 999 cases (larger surveys might require four or even five columns for the *id* number). Notice that the three-column example above requires that we give 'Respondent 1' an *id* code which extends over all of the three columns assigned for the *id* variable. Thus the *id* code for this subject is not 1 but 001.

Column 4 might contain the data for the variable *sex*. In this example, if 1 has been used as the code for males then we could see from the dataset that subjects 1, 3, and 6 are male and that subjects 2, 4 and 5 are female.

Columns 5–6 might represent the variable *age* – the age of the respondents in years. If the survey included only children under 10 then a single column might be assigned (or we might then want to specify age in months and use three columns). It's good practice to be generous in allowing column assignment, although over-generosity may lead to the tedious task of entering lots of unnecessary 0 values.

Columns 7, 8, 9 and **10** contain single-column variables. Whatever variables have been coded in these columns, it appears from our sample matrix that the responses of respondents 1 to 6 have all been coded within the range 1 to 4. Let us suppose that the data in column **7** represent responses to the question **'How far do you agree that women should have the same rights as men?'**, with 'strongly agree' coded as 1 and 'strongly disagree' coded as 5. We could name this variable *sper* and later provide it with a label 'Support for equal rights'.

At this stage we can begin a cursory examination of the data. Notice that none of our subjects (out of our sample of only six) strongly disagreed with this statement about equal rights – there is no code 5. The two respondents who 'disagreed' with the statement (code 4) were men (coded 1 on *sex* – column **4**) and all three women 'strongly agreed' with the statement.

In drawing any conclusions from our data, however, we might wish to draw attention to the fact that this is a young adult sample. From the information in the *age* columns (**5–6**) we can calculate that the average age overall is 21.0 years. Looking at the *age* columns in association with the *sex* column (**4**) we can also calculate that the average age of the men (coded 1) is 21.66 years and that the average age for the women is 20.33 years.

This kind of cursory examination of the data (sometimes referred to as 'eyeballing') is interesting, and we've already begun to think in terms of:

frequencies	there are three men and three women
averages	the average age is 21
differences	the average age of the men is 21.66, whereas the average age of the women is 20.33; also, the women are more supportive of 'equal rights' than the men
associations	it looks as if the variable *sex* may be related to *sper* – support for equal rights.

Such analyses lie at the heart of what SPSS/PC+ is designed to do (more accurately, less tediously, and with much larger datasets). Thus to explore such aspects of our data further, with a much larger sample and using more complex statistical procedures, we need to 'get into' the SPSS/PC+ system. SPSS/PC+ can handle many hundreds of cases and up to 500 variables. Furthermore, SPSS/PC+ will produce neat labelled tables which will enable us to examine and report the findings of our projects in an appropriate and thorough fashion.

Now that you have worked through this chapter you should be able to:

- devise a structured questionnaire
- examine any questionnaire and analyse it fully in terms of the variables it includes
- devise a coding system for each variable
- transform completed questionnaires into a data matrix.

Although this presentation of SPSS/PC+ focusses on data derived from questionnaires, the system can also be used to analyze data obtained from other kinds of study, including marketing projects, demographic studies and psychological experiments. In some cases there will be little need for an elaborate coding system. Each item of information collected from all subjects (or about each company, etc.) can be regarded as a separate variable. The rule for formatting the dataset so that single- or multiple-column slices contain data on a particular variable still holds. Information on the use of decimal data (e.g. 5.328, 66.23) and alphanumeric data (i.e. data coded in terms of letters as well as numbers) will be provided later (in Chapter 16).

3 *SPSS/PC+ and the Computer*

In this chapter we will begin to use the microcomputer to start our interaction with the SPSS/PC+ system.

SPSS/PC+ runs on microcomputers which are 'PC compatible' (a high proportion of personal computers are of this type). They store information on a **hard disk**, an integral part of the machine which can store a huge amount of information (in what are known as **files**). Slots at the front of the machine allow the user to insert 'floppy disks'. Such disks, which are either 5.25 inches or 3.5 inches in diameter, have much less storage capacity than the integral hard disk but can store substantial amounts of information. Files can be copied between the hard disk and the user's disks which are inserted into the slot labelled *Drive A* and, if the machine has two diskette slots, the slot labelled *Drive B*.

Linked to the base unit of the microcomputer is a keyboard which enables the user to enter information into the computer, and a display screen (or visual display unit, VDU). In most cases a printer will also be linked to the machine. Together these units make up the **hardware** of the computer system. **Programs** that run on the computer are referred to as **software**, SPSS/PC+ being one example of a powerful program. Other programs include word processors and a number of programs which control the basic operation of the computer system (some of which are known collectively as the **disk operating system**, or **DOS**).

3.1 The computer keyboard – a guided tour

Computer keyboards come in a variety of designs, and the way in which the computer responds when a particular key is pressed may be changed, in many different ways, by instructions given in programs.

Most keyboards will have the following features, although the exact position on the keyboard, and the labelling on the individual keys, will depend on the particular design.

Apart from the basic alphabetical and numerical characters, and other symbols familiar on typewriters, computer keyboards have many extra keys. The function of this section is to familiarize you with the use of the some of these extra keys, and to show how their use is represented in this book. First, though, we consider a very familiar keyboard action.

The 'space bar' is at the bottom of the keyboard (nearest the user). In this book, an instruction to press this key will be written <SPACE>. To comply with this instruction you therefore simply press the 'space bar' once. You would *not* type S, P, A, C, E or either of the symbols '<' or '>'.

The use of the 'return' key, sometimes labelled 'Enter' and often printed with a bent arrow symbol, will be indicated with <RETURN>. This does not mean that you type the word. Simply press the <RETURN> key once.

The use of 'delete' keys (usually labelled "Del") will be indicated by **<Del>**. A delete key without an arrow, or with an arrow pointing to the right is a 'forwards delete' key. It operates at the 'cursor' (usually a flashing line on the screen which marks the current position at which information can be written) in the following way. When this Del key is pressed briefly, text to the right of the cursor is 'pulled from the right' and deleted one character at a time. Holding this Del key down will cause whole strings of characters to be deleted. The cursor position stays the same throughout. A Del key labelled 'Bck Del', or '<-Del' or simply '<–' deletes text to the left in the following way. When the key is pressed briefly, the cursor moves one column to the left, deleting any character that previously occupied the column and 'pulling' any text to the right of the cursor leftwards.

On the right hand side of the keyboard there is a bank of keys each with a number and an alternative function indicated (including arrows pointing up, down, left and right). This 'keypad' can be operated in one of two modes, either as a 'number pad' or as a 'function pad'. A key labelled "Num Lock" (or alternatively, 'Func Lock') will act as a 'toggle switch', such that one press will change the keypad from one mode of action to the other, and a second press will change it back again. A light on the 'Num Lock' key will indicate which of the modes is operative at any time. If the light is *on* on the 'Num Lock' key, then the keypad will act as a 'number pad' and may be used to type numbers. When this keypad is in 'function mode', pressing the various keys will cause the cursor to move on the screen (typically, within a file containing text or numbers). A single press of the left arrow key will cause the cursor to move one column to the left, for example, and a single press of the up arrow key will cause the cursor to move one row up. Apart from the arrow keys there may be facilities for moving the cursor immediately to the top of the screen (**<Home>**) and to the bottom of the screen (**<End>**), and for the current 'page display' to be superseded by a display of the following 'page' (**<Pg Dn>**) or the preceding 'page' (**<Pg Up>**).

The movement of the cursor may also be controlled from another, smaller, keypad of four keys each marked with an arrow pointing either left, right, up or down.

Of great importance in using the SPSS/PC+ editor program **REVIEW**, are the **function keys** labelled **F1** to **F10** (there may be more, but these are never used in SPSS/PC+). They may be on a keypad to the left of the keyboard, or they might be ranged across a line at the top of the keyboard. Instructions to use one of these keys will be indicated by **<F1>**, **<F2>**, etc. Sometimes these function keys need to be pressed briefly while another key (such as the 'Alt' key, see below) is held down, and in other cases the press of a function key needs to be followed by another key press to obtain the desired effect. The effects of the functions keys when used in conjunction with other keys are considerably different from their effects when pressed alone.

'Shift keys' are normally used to type upper-case – capital-letters – and the various symbols which appear above the numbers at the top of the keyboard. The shift keys often have upward-pointing arrows printed on them, and they have an effect only when pressed at the same time as another key. Similarly, the 'control key' (usually labelled 'Ctrl') only has an effect when it is pressed at the same time as another key. We will write **<Ctrl>** to indicate that the control key should be pressed. Thus the instruction **<Ctrl-C>** would mean 'press the control key in association with the c key'. The command **<Ctrl-C>** is often useful in interrupting an operation which is underway. Another way of bringing about a 'break' or exit from a current process or level of operation is to press the 'escape' key (usually labelled 'Esc'). An instruction to press this key will be written **<Esc>**.

The 'Alt key' (usually labelled 'Alt') is also similar in that it has an effect only when pressed in combination with another key. We will use the symbol **Alt-** to indicate that the Alt key should be pressed. Thus the instruction **<Alt-F10>** would mean 'press the Alt key in association with the F10 key'.

Within **REVIEW** there are many cases in which operations are commanded by a succession of two key strokes, the first of which involves pressing one of the special function keys. Thus the two keys are *not* pressed simultaneously but in succession. This is indicated by a pair of symbols written together, but each within pointed brackets. For example **<F9><w>** means 'press the F9 key, release it, and *then* press the w key'.

3.2 Drives, directories and files

In your interaction with the microcomputer you will be writing, reading and using various types of SPSS/PC+ **file**. Files can contain data, data descriptions, commands or any combination of these. Within files, information is stored in one of two ways – either in **binary** form (i.e. as 'bits' of information) or in **ASCII** form. An ASCII file is one that includes only characters conforming to the American Standard Code for Information Interchange standards. In practice this means that the file contains only plain text and numbers, without the additional 'command' characters used, for example, in word-processor programs and database management programs. The binary form is more efficient in some ways but ASCII files have the advantage that they can be easily edited and printed. Eventually we will learn how to instruct the machine to create binary files, but the first files we will create will be written to disk as ASCII (or text) files.

Because the hard disk is capable of storing many hundreds of different files, it is organised into different logical sections. Computers with smaller hard disk capacity have just one 'drive' on the hard disk (*Drive C*), whereas those with greater hard disk capacity might also have one or more additional drives (*Drive D*, *Drive E*, etc.). Each drive on the hard disk is divided into a number of **directories** each of which contains various files. Thus on my own computer I use *Drive E* to store all information relating to writing. I have separate directories for each writing project, and within the directories I have separate files relating to each paper or book chapter. I am now writing this in *Drive E*, I have labelled the directory "SPSSBOOK", and the file has been named "2spss3" (i.e. second edition of *A Guide to SPSS/PC+*, Chapter 3). The full specification for the present file is therefore:

```
E:\spssbook\2spss3
```

For each drive (C, D, etc.) there is a basic **root directory** from which the other directories in that drive branch. Thus the hard disk may have several drives, each drive has several directories, and each directory contains several files. With these various branches and sub-branches, the logical structure of the hard disk is rather like a tree. Finding one's way through the tree to access a particular file could be a problem (particularly when someone else has created the logical structure of the hard disk) but fortunately **path commands** can be issued to the machine so that files stored within one directory can be accessed from other directories, including those on other drives. The computer you will be using for SPSS/PC+ analysis is likely to have been set up with a directory named "SPSS" containing all of the SPSS/PC+ system programs, and it is probable that a suitable path command, allowing SPSS/PC+ to be accessible automatically from any directory, has been entered into a special file (the **autoexec.bat** file) that is executed each time the computer is switched on.

If this *is* the case then you will be able to operate SPSS/PC+ from any directory simply by issuing a simple command. In some cases the machine will have been set up so that when it is switched on, and after various initial programs have been run, the user will be presented with a **Menu** from which to choose one of a number of programs to run. In such cases, simply pressing one **key** (perhaps S for SPSS/PC+) may be sufficient to start to run the SPSS/PC+ program.

It would be impossible to cover here all of the possible 'front end' arrangements that might be in operation on the computer you are using. There are very many different ways of setting up a computer so that SPSS/PC+ can be used. On the first occasion you use the system you may need to ask someone who is familiar with using SPSS/PC+ *on that machine* how to access the program. Thereafter you should experience no difficulty in running SPSS/PC+, and once the program is running you will be able to follow all of the instructions given in this book.

If you are installing SPSS/PC+ yourself, of course, then you are the one who will decide how it will be accessed. Installation is a very simple process; simply place the first SPSS/PC+ diskette into *Drive A* or *Drive B*. Log to that drive (by issuing **<a:>** or **<b:>**, as appropriate) and type "Install" followed by <RETURN>. Then follow the on-screen instructions about when to change disks, etc. When you have completed the installation process the SPSS/PC+ system will be loaded into a directory named "SPSS" on *Drive C*. A suitable path command should now be added to your 'autoexec.bat' file so that SPSS/PC+ will be accessible from any directory (consult your **DOS** manual if you need help with this).

To use the computer efficiently it is best to create your own directory and to work from this. Thus your own files of data, etc. do not need to be stored in the SPSS directory itself. Before describing how such a personal directory can be created, however, we need to understand the different levels of program which will be involved in our use of SPSS/PC+ on the microcomputer.

Using SPSS/PC+ involves dealing with three levels of program:

- **DOS** – the 'disk operating system' – a control program
- SPSS/PC+ itself
- **REVIEW** – the SPSS/PC+ editor program

3.3 The three program levels

Level 1: DOS (the disk operating system)

DOS consists of a group of programs which enables the user to manipulate information stored either on the hard disk or on diskettes. **DOS** can be used to create, copy and print files, to run programs, and to organize and switch between directories.

Level 2: SPSS/PC+

SPSS/PC+ contains all of the data-handling and statistics programs of the SPSS/PC+ system.

Level 3: REVIEW

This is an editor program within the SPSS/PC+ system which allows the user to create and edit files. **REVIEW** is SPSS/PC+'s own editor program. Other editor programs (including some word-processing programs) can also be used to create files for use by SPSS/PC+ (a general description of how SPSS/PC+ files can be created and edited using alternative programs is given in *Appendix B*). However, the fact that **REVIEW** is fully integrated with SPSS/PC+ and can be entered directly during an SPSS/PC+ session, means that it is generally the most efficient editor to use. It is recommended that you spend the little effort necessary to learn how to deal with **REVIEW** (a detailed presentation begins in the next chapter).

REVIEW can edit files created by other editors, but only if they are ASCII files. The files normally produced by word-processors contain information other than that expected by SPSS/PC+, although some word-processing programs do allow the user to save a file in ASCII form.

One way to find out whether a file is in ASCII form is to try to display it on the screen using the **DOS** 'type command', using the format:

```
C>type myfile <RETURN>
```

If all of the characters on the screen are alphanumeric, then the file is probably in ASCII form.

A typical SPSS/PC+ session

If **REVIEW** is being used as the editor program, a typical session will involve entry through **DOS** to SPSS/PC+, followed by a series of changes back and forth between **REVIEW** and SPSS/PC+. After issuing the command **FINISH** to SPSS/PC+ the user will be returned to **DOS** and may then wish to copy files that have been produced during the session, using **DOS**, to a diskette in *Drive A* or *Drive B*.

The next section describes how you can move between the different levels of program – **DOS**, SPSS/PC+ and **REVIEW**.

3.4 Moving between program levels

Entering DOS

DOS is usually in operation as soon as you switch on the computer, so 'entering **DOS**' usually involves simply switching on the machine (at this stage you should make sure that there is no floppy disk in either *Drive A* or *Drive B*). When internal checks have been made (this may take up to 30 seconds) the **DOS** 'prompt' (typically, **C>**) will automatically appear. **DOS** is now in operation, waiting for a command. You then type your command on the screen at the current **cursor position**.

You will (automatically) be in the **root directory**.

If you have previously created a directory for yourself then you can enter your directory with a 'change directory' (**cd**) command, followed by the appropriate directory name. After typing any instruction into the computer you should press the <RETURN> key (also referred to as the 'Enter' key; this key may be indicated on the keyboard with a bent arrow). This **enters** the instruction into the system. Thus:

```
C>cd\johndoe                <RETURN>
```

If you have NOT previously created a directory for yourself you can create one with a 'make directory' (**md**) command, followed by the name you wish to give to your directory. The name can be up to 8 characters long. Thus:

```
C>md\samspade               <RETURN>
```

(For commands given to **DOS** – as for SPSS/PC+ commands – no distinction is made between upper-case letters – i.e. capitals – and lower-case letters).

Now change directory (**cd**) to enter your newly created directory:

```
C>cd\samspade               <RETURN>
```

(If at any stage you need to return to the root directory this can be done by keying **cd** and <RETURN>).

Assuming that the system has already been set up appropriately for the easy use of SPSS/PC+, you will now be able to access the SPSS/PC+ system from your own directory by following the instructions below. (If access proves difficult, consult somebody who is familiar with how the particular computer has been set up. If you are installing SPSS/PC+ for the first time on a machine of your own, follow the installation instructions and the instructions in your **DOS** manual on how to create paths, etc.).

Entering SPSS/PC+

It is assumed that SPSS/PC+ will have been pre-loaded onto the hard disk of your machine. Although the SPSS/PC+ package will be stored on the hard disk in a directory named "SPSS", a suitable path command should automatically be in operation so that the program is accessible from any other directory, including your own directory.

Once **DOS** has been loaded and you have entered your own directory you are ready to enter SPSS/PC+. Assuming that a suitable path command is in operation on the machine, simply type "spsspc" (upper or lower case letters) and press <RETURN>.

```
C> spsspc                   <RETURN>
```

The SPSS/PC+ logo will appear, followed by the opening display. On later editions of the program this is a rather complex entry display with a screen split into several sections.

It is strongly recommended that you *bypass* this (the reasons for this are outlined in *Appendix A – Notes for Trainers*). To bypass this display press down the key marked "Alt" and, while holding this down, briefly press the key marked "F10" (using the convention employed throughout this book, this combination of key presses is shown as **<Alt-F10>**).

This takes you out of the opening display and presents you with a clear screen containing only the 'command prompt' which, throughout each SPSS/PC+ session, will tell you that SPSS/PC+ is awaiting a command. The command prompt looks like this:

```
SPSS/PC:
```

Entering REVIEW

To use the editor, simply type, at the SPSS/PC+ prompt, **REVIEW** (again upper- and lower-case letters are treated as equivalent) followed by the name of a file to be edited (the file name must be included within apostrophes – *double* quotation marks). End with a period (.) and then press <RETURN>. For example:

```
SPSS/PC: review "trial.dat".      <RETURN>
```

Note that the period (.) should be included and is always used to signal the end of an SPSS/PC+ command.

If, in the above example, "trial.dat" was the name of a file that had already been written, the **REVIEW** command would present a copy of the file on the screen, ready for editing. If, however, no file with that name exists, you will be presented with a blank screen area on which you can write the information which will later be stored on the disk in a file called "trial.dat".

Try typing a few letters and numbers. Press <RETURN> to see how this moves the cursor to a new line, and then use the cursor control keys to move about the text you have created. Try also to produce the double quotation marks symbol (") on the screen. This might involve simply pressing the shift key and the key which has, as the upper symbol, the appropriate (") marking. However, a keyboard controlling program may be in operation (which has changed this. You may find that pressing the shift key together with **<">** results in the symbol '@' appearing on the screen, for example. If this is the case, experiment until you find the keys that lead to a " symbol appearing on the screen (you may find that 'shift' and **<@>** will achieve this).

Getting back

For now, let us abort the file we have been creating. Do this by pressing **<Alt-F10>**. You will then be presented with the warning message **'Unsaved changes in "trial.dat" – ok to exit'** followed by a highlighted 'N'. Simply type **<y>** (for 'yes') and press <RETURN> and you will exit from **REVIEW** and be presented with the SPSS/PC: command prompt. SPSS/PC+ is awaiting your next command.

To exit from SPSS/PC+ we use the command **FINISH.** (again terminated with a period – **.**).

```
SPSS/PC: finish.
```

The **DOS** prompt (**C>**, or something similar) will then re-appear. You are now back in **DOS** and in the directory from which you entered SPSS/PC+.

You have now learned how to move back and forth between the three levels of program you will use in a SPSS/PC+ session.

Sooner or later you will make a mistake in the syntax of one of the commands you issue to SPSS/PC+. The system will inform you of this, generally by displaying an error message with an explanation of your mistake. This is one of the most useful features of the SPSS/PC+ system. A very common error is to omit the terminal period (**.**) of a command. This will not produce an error message, but SPSS/PC+ will present a 'continuation prompt' (**:**). If you press <RETURN> in response to this, SPSS/PC+ will then accept the previously issued command.

Interacting with SPSS/PC+

From **DOS**, get back into SPSS/PC+ once again. Type "spsspc" (followed by <RETURN>). After the logo display has disappeared, and the entry display appears, press **<Alt-F10>** to get to the SPSS/PC: prompt.

Many different types of command can normally be issued in response to this prompt, but since we have not yet created any files containing information to be analyzed we are somewhat restricted at this stage. We can, however, explore some **operation commands** which allow the user to obtain useful on-screen information and to change some of the characteristics of the system.

Two such operation commands will be examined at this stage – **SHOW** and **SET**.

3.5 The SHOW and SET commands

The **SHOW** command produces a display of the current state of many aspects of the SPSS/PC+ system. The **SET** command allows these aspects to be changed.

The **SHOW** display is produced merely by entering 'show.' in response to the SPSS/PC+ prompt:

```
SPSS/PC: show.
```

No further specifications are necessary for or applicable to the **SHOW** command.

Much of the information displayed on the screen in response to the **SHOW** command can be ignored at this point, but let us look at just some of the items and consider how certain aspects of the general operation of the SPSS/PC+ system can be changed by an appropriate use of the **SET** command.

Look, in particular, at the current ON/OFF status of the **Printer** and of two other aspects – **More** and **Beep**.

These (and the other items included in the **SHOW** display) can be changed by issuing a suitable **SET** command in response to the SPSS/PC+ prompt.

Thus, if the **printer** is currently off, it may be switched *on* with the command:

```
SPSS/PC: set printer = on.
```

Or, if the printer is on, and you wish to switch it *off*, you would use:

```
SPSS/PC: set printer = off.
```

These commands of course refer specifically to the interaction of the SPSS/PC+ system with the printer. Obviously the printer itself must be switched on if it is to operate at all.

Notice the spacing used in either of the above commands. SPSS/PC+ is very tolerant of 'extra' spaces between words and symbols. Both:

```
SPSS/PC: set printer=off.

SPSS/PC: set printer  =          off.
```

for example, would be accepted by the system. However, there must be at least one space between the command word and the subcommand (and between variable names, etc., when these are used) and there must not be a space *within* a single term. Thus, neither of the following would be acceptable to the system:

```
SPSS/PC: setprinter = off.

SPSS/PC: set pr inter = off.
```

The output from certain procedures, including statistical analyses, often extends over many 'pages' (and thus many screen displays). Normally, SPSS/PC+ presents the first page of information on the screen, at the same time issuing a **More** message at the top right-hand corner of the screen and causing a high-pitched **Beep** to sound.

This allows the user to read one screen of output before the next one is presented. The initial display remains until, by pressing any key, the user brings the next page of information to the screen.

Sometimes, however, the user will not wish to read a screen display of the output from an analysis but would prefer, perhaps, to read the output in printed form at some later time (a complete record of all output produced during a session is automatically saved by SPSS/PC+ in a file named "SPSS.LIS"). With the **More** function *on*, the user would need to wait at the machine and repeatedly press a key to move the output on to the next page. With the **More** function set to *off*, however, there will be no delay and no need to press a key every time a new screen is presented. Output images will flash past quickly until the procedure has been completed and the SPSS/PC: command prompt returns.

More can be switched *off* by using the **SET** command:

```
SPSS/PC: set more = off.
```

and can be switched on again:

```
SPSS/PC: set more = on.
```

The audio signal **Beep** has two functions. A high-pitched beep merely alerts the user to the fact that the current display is being held and that a key-press is necessary if further output is to be displayed. A low-pitched beep signals an error.

If **More** is switched *off* then the high-pitched **Beep** that normally accompanies it is not heard. But if the user wishes to switch the **Beep** *off* while **More** remains *on*, this can be achieved by issuing the **SET** command:

```
SPSS/PC: set beep = off.
```

As you are introduced to new features of the SPSS/PC+ system you will come to understand more of the items in the initially bewildering display elicited by the **SHOW** command. You will also learn when it is appropriate to use a **SET** command to change particular aspects of the system. Generally, it is not essential to change *any* aspect in order for SPSS/PC+ to run successfully because the settings which are provided automatically allow the system to function perfectly well (although to obtain a direct printout of your output you will have to **SET** the printer *on*). The automatic settings provided by the system (including the names provided for certain files, the colours used on the screen, etc.) are known as 'default' settings. They operate by default i.e.

unless the user deliberately changes one or more of them with a **SET** command. A clever use of **SET** commands, however, can add flexibility and allow the user to work the system at maximum efficiency.

Having now learned how to switch between levels of program (**DOS**, SPSS/PC+ and **REVIEW**) and how to interact with SPSS/PC+ directly – using the **operation commands SHOW** and **SET** – it is time to consider how **files** can be created and stored on disk. We will start, in the next chapter, by using **REVIEW** to enter a dataset into a **data file** which can be written to the disk for later analysis by SPSS/PC+.

4 *Creating a Data File*

In this chapter we will present some basic guidelines for using the SPSS/PC+ editor program **REVIEW**. We will use this editor to enter data and to create a **data file** which can be stored on disk.

4.1 Using REVIEW

REVIEW is the SPSS/PC+ editor program. Because of its power and flexibility, **REVIEW** is somewhat complex, and in this chapter only some of the many features of **REVIEW** will be introduced. Enough information will be given, however, to allow you to enter data which have been coded onto questionnaire forms (or onto **data sheets**) into a **data file**. Further aspects of **REVIEW** will be introduced in later chapters.

Some of the **function keys** labelled 'F1' to 'F10' are used for many of the **REVIEW** editing operations (an instruction to press one of these keys is shown as **<F1>**, **<F2>**, etc.).

These function keys have a different effect when they are pressed briefly while another key (such as the 'Alt key') is held down. Thus the instruction **<Alt-F10>** (i.e. both keys indicated within a single pair of pointed brackets) would mean 'press the F10 key in association with the Alt key'.

In other cases a particular operation can be executed by pressing one of the function keys *followed by* another key (usually a letter). In this case two keys are not pressed simultaneously but in succession. This is indicated by a pair of symbols written together, but each within its own pair of pointed brackets. For example **<F4><d>** means 'press the F4 key, release it, and *then* press the d key'.

If you are still in **DOS** (i.e. if the screen presents you with a **DOS** prompt – usually **C>** or **D>**) then type **spsspc** <RETURN>. The screen will display the SPSS/PC+ logo, followed by the opening screen. Bypass this in the recommended way with **<Alt-F10>** and the SPSS/PC+ command prompt (SPSS/PC:) will appear.

Respond to this prompt by typing "review" (it doesn't matter whether you use upper-case letters, lower-case letters, or even a mixture of the two) followed by the name of the data file *to be created*. File names often include a **file extension** of three letters following a period. Thus a word-processing file which contains text might be given the name "essay.txt", where "txt" is the extension. It is useful to think of files as having a 'forename' which comes before a period (.) and a 'surname' which comes after. The forename can be up to eight characters (letters and/or numbers) in length. The three letter surname can be chosen to best describe the nature of the file. Thus a data file may be given the surname "dat". Throughout this book, files which contain only data will be given names with this extension (thus "trial.dat", "gender.dat", etc.).

Before creating a data file that contains real data, let us explore some editing functions with a file of 'dummy data'. First use **REVIEW** to create a data file called "dummy.dat" by entering the instruction:

```
SPSS/PC: review "dummy.dat".  <RETURN>
```

Notice that because "dummy.dat" is the **name** of a file (rather than a direct command like **SHOW** or **REVIEW**, it is typed within quotation marks. The final period (.) is used to signal the termination of the **REVIEW** command.

If the command has been typed correctly a screen will be presented on which we may start to enter data. Now use the number keys just above the 'Q W E R T Y, etc.' keys to enter a row of numbers (any numbers will do) on the top line of the screen. When you have filled about half of the line, press <RETURN> and enter a second row of numbers. When the numbers extend as far as those of the first row, press <RETURN> again and continue to enter numbers in this way until you have five or six rows completed. This can act as a substitute for a real data matrix, while we explore a few useful features of the editor program **REVIEW**.

You may have noticed, as you entered the dummy data, a number of messages at the bottom of the screen. The file name "dummy.dat" will be displayed, as well as a message "Std Menus" which need not concern us here. Two other items, which *are* important, are a number at the extreme right of the display, which indicates the current cursor column position (this number therefore changes as we move the cursor or type data), and 'Ins'. Press the **<Ins>** on the keyboard once and the 'Ins' message on the screen will disappear. Try using the <RETURN> key and you will find that the cursor will not now move beyond the last line of text to create a new line, as it would when the 'Ins' message was present. Press **<Ins>** once again and the 'Ins' message will reappear.

When the 'Ins' message is present we know that the editor program is in **insert** mode. Sometimes it is useful to have **insert on** and sometimes it is more useful to have **insert off**. If the insert is *on*, any typed characters will be inserted at the current cursor point. Thus if we have a display:

```
11113333
```

and we place the cursor at the first 3, then typing 2 four times will produce the result:

```
111122223333
```

With the insert *off*, however, the typed characters will **overwrite** any existing characters. The same key presses would thus result in:

```
11112222
```

If a data item has been mistakenly omitted from a data-line, therefore, we could insert the missing character(s) at the correct point with the 'Ins' *on*. If, however, we need to overwrite one or more incorrect numbers then we would do this with the 'Ins' switched *off*. With 'Ins' *on*, pressing <RETURN> will create a new line below the current line.

Now find another key, labelled 'Num Lock'. Press it once and the message 'Num' will appear at the bottom of the screen. SPSS/PC+ starts with 'Num' *off*, so that the bank of keys on the **number pad** (usually on the right hand side of the keyboard) is switched to 'function mode'. The keys can then be used to move the cursor, etc. However, these same keys can also be set to the alternative 'number mode', in which case they are used to enter numbers. This is achieved simply by pressing the 'Num Lock' key once. Pressing it once again restores the 'function' operation of the keys. When the 'Num' message is absent from the **REVIEW** screen the function mode is in operation. Remember that another set of cursor keys are to be found on the keyboard,

With the number pad in **function mode**, one key will act as a **delete** key. Pressing this will delete the character at the current cursor position.

The keys on the number pad, whether they are switched to 'number mode' or 'function mode' should not be confused with the bank of special **function keys** on the keyboard that are labelled 'F1' to 'F10' (there may be more, but these are not used by SPSS/PC+). These function keys have particular uses within **REVIEW**.

Let us begin to explore the further editing facilities of **REVIEW**. After you have started entering a line of data press the F4 key and then the d key. Notice that this has erased the line of your data on which the cursor was placed. This is because the F4 key controls actions to be performed on a single line, and the d that follows instructs the system to "*D*elete" the current line. Move the cursor to some other position within the matrix of dummy data and now press **<F4><u>** (i.e. the F4 key followed by the u key). You will see that the line that was previously deleted has now reappeared at the new position. This is because the F4 key that identifies a particular line for an editing operation, was followed by u for "*U*ndelete". The overall effect of the two actions you have just performed will therefore be to have moved a line of data from one position within the matrix to another.

You might have noticed, as you performed one or both of the actions **<F4><d>** and **<F4><u>**, that immediately after pressing the F4 key a 'minimenu' appeared at the bottom of the screen. Press just the one key, **<F4>** now to see this minimenu.

The first word 'lines' identifies the general nature of the operations available at this stage (here, to manipulate lines). Four alternative operations are presented: "Insert after", "insert Before", "Delete" and "Undelete". "Insert after" means insert a blank line immediately *after* the line with the cursor, and "insert Before" means insert a blank line immediately *before* the line with the cursor. The single capital letter contained within each of the four messages is important because it signifies which letter should be pressed on the keyboard to carry out the particular operation. Thus "Delete" informs us that pressing **<d>** will cause the current line (i.e. the line which the cursor is on) to be deleted, and "insert Before" tells us that **<b>** will insert a line *before* the current line.

You will notice that the first option, "Insert after", is highlighted. This means that this option can be accepted simply by pressing <RETURN>. To accept another option in this way move the highlight to the right by pressing the right cursor key, and then press <RETURN> to accept the highlighted option. Finally, as the highlight moves you will notice that a message to the extreme right also changes (from an initial "Alt-I", through "Alt-B" and "Alt-D" to "Alt-U"). This provides information on yet another way of performing the desired action on a single line. In this case, however, the Alt key must be pressed *at the same time* as the other key, not immediately before. Thus there are three different ways in which, for example, the current line may be deleted in **REVIEW**:

(a) **<F4><d>** : The F4 key may be pressed, and then the d key
(b) **<F4>** : This provides a minimenu, with options. The highlight is moved if necessary, and then <RETURN> is entered to accept the highlighted option.
(c) **<Alt-d>** : The Alt key is held down while the d key is briefly pressed.

As well as performing editing functions on single lines, we can perform such operations on an *area* of the file being created. An area is either a set of whole lines or a 'rectangle', a rectangle being a perfectly rectangular area (or block) consisting not of whole lines but only of a part of each of a number of consecutive lines.

Before operating on a block, we first need to mark it, using **<F7>**. To mark a block of whole lines, place the cursor on the first (or last) line of the intended block, and issue the command **<F7><l>**. You will notice that the line begins to flash and that a message appears: **'Waiting for second block**

marker'. Now move the cursor to any position on the line which is to be the last line of the block and press **<F7>** again. The block has now been marked and the lines on the screen which are included in the block will change to a different shade (or colour). A message will tell you that the block has been marked and also tell you how many lines are included in the block (**'Area marked *nn* lines'**).

An alternative method to that of entering **<F7><l>** is to enter **<F7>** and, when the minimenu appears, to highlight the "Lines" option and press <RETURN> (no alternative 'Alt- ' commands are available for area marking or area manipulation).

To define a rectangle place the cursor at the position which is to mark the extreme top-left corner of the rectangle and press **<F7><r>**. The **'Waiting for second block marker'** message will appear. Now move the cursor to the point which is to be the extreme bottom right corner of the rectangle and press **<F7>** again. The area has now been marked and will be highlighted, and the message **'Area marked *nn* lines'** will appear.

To cancel a marked area at any time, simply press **<F7>** once. The highlighting will disappear and the message **"Area cancelled"** will appear.

When an area is marked it can be manipulated in various ways using **<F8>**. Of interest to us are the three operations 'Copy', 'Move' and 'Delete'. After an area (a block of complete lines, or a rectangle) has been marked, **<F8><d>** will delete it. To copy the block, move the cursor to the position where the copy is to be inserted and enter **<F8><c>**, and to move the block move the cursor to the position to which the point to which the area is to be moved and enter **<F8><m>**.

Many other editing functions are available, as you will see if you enter **<F1><r>**. You will be presented with a display of the various **REVIEW** commands. There are rather a lot. Don't worry about all of the available functions at this stage – many are merely 'short cuts' to effects which can be achieved in other ways. Notice the F4, F7, and F8 items, and the features of these which you have already used. Now press <SPACE> on the keyboard and you will be returned to the file you are editing.

We have been working with dummy data to demonstrate some of the **REVIEW** editing features. Now it is time to begin creating a real data file. Attempt to abandon the current file with **<Alt-F10>** and you will be presented with a warning message: **'Unsaved changes in "dummy.dat" – ok to exit'** followed by a highlighted 'N'. Simply type **<y>** (for "yes") and press <RETURN> and you will exit from **REVIEW** and be presented with the SPSS/PC: command prompt. The file "dummy.dat" will not have been saved, but when we come to enter real data then we will save the file as a data file.

4.2 Entering a data matrix and saving it as a data file

Now begin to enter a real data matrix. You may wish to create a file of data from a study of your own, or you may make a copy of the data from the Gender Attitudes Survey (reproduced as *Appendix E*).

Begin by entering the data for case 1 across the top line of the screen in the form of a long string of numbers, without spaces (unless, of course, you have chosen to use spaces to separate the data relating to each variable). If the data from a single case extend over more than one 80 column line, press <RETURN> at the end of the first line and continue entering the data for that respondent on column 1 of a second line (you may have decided to indent the second and subsequent lines (records) for a case by leaving a set number of columns blank before entering data).

Continue in this way for all of the data relating to each case in your study. If you notice at some point that the number you have just typed is a mistake, place the cursor over the mistaken number and press the delete key (**<Del>**). If you notice a mistake when you have already entered further data then move the cursor to the appropriate position (taking care that **Ins** is *on* or *off*, as appropriate) and insert any missing numbers, or overwrite or delete errors.

When all of the data have been entered you will be able to save the data matrix as a data file. This will be stored on the disk for later use by SPSS/PC+.

4.3 To save the file

To save the complete contents of the file you have created, you use the command **<F9><w>**. This instructs the program to 'write the *W*hole file to disk'. You will be presented with a prompt requesting that you enter a name to be given to the file which will be saved. Initially, the prompt is followed by the name originally given to the file, and if you wish to accept this name for the file you have created you simply need to accept the filename by pressing <RETURN>. If you decide to change the name, type in another suitable name (you do *not* need quotation marks) and press <RETURN>. The data matrix will then be stored as a data file on the hard disk.

When this prompt (or any other **REVIEW** prompt) is presented, if you need advice on what to do, pressing **<F1>** will produce an explanatory message. Also, if you decide that you wish to abort the file naming operation and to return to editing, you may do this by pressing **<Esc>**.

If you do save the file at this stage, and you use a filename that already exists, the existing file of that name will not be deleted but will be saved as a backup file, with the file extension (surname) changed to '.bak'.

Having successfully saved a complete **REVIEW** file, you can now exit from **REVIEW** by issuing the command **<Alt-F10>**. You will be presented with the SPSS/PC: prompt.

Congratulations – you have created a data file which is now stored in your directory of the hard disk. At this stage you might like to copy the file to a floppy disk and to obtain a printout of the contents of the data file you have created.

To do this you first return to the disk operating system (**DOS**). Enter **finish.** to end the SPSS/PC+ session and return to **DOS**.

```
SPSS/PC: finish.
```

You will then be presented with the **DOS** prompt (**C>**).

4.4 To copy and print the file

We are now going to use **DOS** to make a copy of our file and to obtain a printed copy of the file contents. It is always good practice to make at least one disk copy and one printed copy ('hard copy') of each of your files. The original may be accidentally erased, but this will not present a serious problem if you have made a 'backup copy'.

To make such a copy, first place a suitable floppy disk in *Drive A* or *Drive B*. If the disk is a new one, you will need to format it before you can copy a file to it. To do this enter, at the **C>** prompt, the command:

```
C> format a:       <RETURN>
```

or

```
C> format b:       <RETURN>
```

(depending on which drive you have inserted your disk into). The formatting process takes a little while, and when the process is completed the question **"Format another? (y/n)"** will appear. Type n, followed by <RETURN> and the **C>** prompt will reappear.

For a previously used disk, it is useful to know what files are already on the disk, and how much space is available. Enter the command:

```
C> dir a:       <RETURN>
```

or

```
C> dir b:       <RETURN>
```

as appropriate, and you will be presented with a list of the files already written on the disk and a note of the unused space available. The space will be indicated in terms of 'bytes free'. Divide this number by 1000 (or simply ignore the last three digits) and this will tell you how many 'kilobytes' (*k*) are free. If you have 100*k* of space, or more, you should have no trouble in writing a data file (and, later, a data definition file) to the disk. The Gender Attitudes data file occupies only 4*k*, and the corresponding data definition file occupies 3*k*, but output files, which you may later wish to copy to the disk, may be considerably larger.

To copy the data file you have written (let's say this is "gender.dat") to *Drive A*, for example, we would issue the command:

```
C> copy gender.dat a:     <RETURN>
```

This will write a copy of the file (with the same name) to the disk in *Drive A*. (If you wish to check that the copy has been made, you may issue another 'dir a:' command).

Remove the disk from the drive, take it with you and keep it in a safe place, in case when you return to the computer you find that some electronic or human mishap has led to your original file being deleted.

Before we obtain a printed copy of the contents of the data file, let us display the data on the screen. Enter the command:

```
C> type trial.dat          <RETURN>
```

and the contents of the data file will be displayed (if you have a long file, only the bottom lines will remain visible). An extension of this command:

```
C> type trial.dat>prn
```

will provide a printed copy (provided, of course, that you have a printer, that it has a power supply, that it has been switched on mechanically, and that it is connected appropriately and is 'on-line' to the computer).

The printed record can be examined for errors. It is worth checking that your data block is perfectly rectangular. If, on the right side of the data matrix, one or more lines stop short of the normal edge of the matrix, or if one or more lines protrude beyond this edge, then you have either left out data digits or inserted extra ones somewhere along the line. Refer back to your data source and **REVIEW** the file to correct the errors.

4.5 Entering REVIEW directly from DOS

Although **REVIEW** is usually entered via the SPSS/PC+ system you can enter the program directly from **DOS**. This is done by typing, at the **DOS** prompt, 'spsspc/re', followed by a space and then a filename. For example:

```
C>spsspc/re mystudy.dat
```

If the file you wish to edit is on a disk in *Drive A*, for example, you may first copy it to your directory on the hard disk. From within your directory, you would enter the appropriate **DOS** copy command at the C> prompt. For example:

```
C>copy a:mystudy.dat
```

Alternatively, you could work directly with the file on the floppy disk, but to do this you would have to refer to the drive in your specification of the filename to be edited with **REVIEW**. For example:

```
C>spsspc/re a:mystudy.dat
```

The file is then edited in the normal way, using the usual **REVIEW** editing commands. The file may then be saved. Following this, pressing **<Alt-F10>** will return you directly to the **DOS** C> prompt.

4.6 Looking forward

Creating a data file with **REVIEW** is an important step towards using SPSS/PC+, but so far the file is just a long list of numbers. These must now be identified to the system as **variables**. The variables must be named, and the position they occupy within the data file must be fully specified. Additionally, the system should be informed of which values signify 'missing' data, and labels may be supplied for particular variables and the values associated with them.

Such information is generally specified in a **data definition file**. Such a file is also created using an editor program. In Chapter 5 we examine how a data definition file may be created using **REVIEW**.

Table 4.1 overleaf summarizes the **REVIEW** commands encountered so far.

Table 4.1

Command	**Key(s)**
REVIEW help (This produces a screen display of the various commands used by **REVIEW** – it can be 'called up' at any stage and the screen display will not interfere with the file you are editing. You can return to continue editing by pressing <SPACE>).	**<F1><r>**
Delete a line	**<F4><d>**
Undelete a line	**<F4><u>**
Insert a line after current cursor position	**<F4><i>**
Insert a line before current cursor position	**<F4><b>**
Mark area of lines then move cursor to end line and	**<F7><l>** **<F7>**
Mark rectangle area then move cursor to end line and	**<F7><r>** **<F7>**
Unmark area	**<F7>**
Delete area	**<F8><d>**
Copy area	**<F8><c>**
Move area	**<F8><m>**
Write a whole file to disk	**<F9><w>**
EXIT from **REVIEW**	**<Alt-F10>**

5 *Creating a Data Definition File*

Creating a data file, containing only the data, is just a first stage towards using SPSS/PC+. The data must be *defined* (and are also usually more fully *described*) before they can be used. The data are defined in a **data definition file**. This usually contains the following types of command:

DATA LIST: a list which names the variables and indicates where they may be found within the data matrix
VARIABLE LABELS: optional variable *labels* should not be confused with the variable *names* specified on the DATA LIST
VALUE LABELS: optional value labels for some or all of the values for each variable
MISSING VALUE: information about which value, if any, represents missing data for each variable

Of these commands only the **DATA LIST** command is *essential*, and if you did not wish to label your variables, and if you had no missing data, then your data definition file might consist of just the **DATA LIST** command. For example:

```
DATA LIST FILE = "TEST.DAT" /id 1-3 sex 4 age 5-6 IQ 7-9.
```

Here the system is informed that data are to be found in a file named "test.dat" and represent four variables (*id*, *sex*, *age* and *iq*) which have been written sequentially for each case in the column(s) specified.

If other commands are to be included in the data definition file (as is usually the case) then these commands must *follow* the **DATA LIST** command. There are few rules in SPSS/PC+ about the order in which the commands are entered into a data definition file, but a variable does need to be declared in a **DATA LIST** command *before* any labels are provided for it or a missing value specified for it.

A glance at *Appendix F* at this stage will give you some idea of what a complete data definition file looks like. This is the data definition file used to define and label data from the Gender Attitudes Survey. Don't try to understand all of the commands within this file at this stage, but notice that the rather long file contains only four commands, one specifying the variables and their location within the dataset (the **DATA LIST** command), two specifying labels for some of the variables and some of the values that the various variables may take, and one specifying values assigned to 'missing data'. Each of these types of command will be explained fully within this chapter.

Data definition files can be created using **REVIEW** or another suitable editor program. We will use **REVIEW** to create a simple SPSS/PC+ data definition file which can be used with a data matrix which has been saved in a previously written data file (see Chapter 4). We will also use **REVIEW** to re-edit the file if any errors emerge, and to correct any data-entry errors which are detected in a printout.

To enter **REVIEW**, we first enter SPSS/PC+. We do this simply by responding to the **DOS** prompt with "spsspc":

```
C> spsspc   <RETURN>
```

The SPSS/PC+ logo will appear, followed by the menu and split-screen display. As usual, we will bypass this display by issuing the command **<Alt-F10>** and we are then presented with the prompt (SPSS/PC:) which acts as our starting point for issuing any SPSS/PC+ command.

In this case we issue a **REVIEW** command which includes an appropriate name for our data definition file. It is a good idea to use the extension (or 'surname') **def** for data definition files. For example:

```
SPSS/PC: review "gender.def".
```

The system will respond by presenting a screen on which we can start to enter information. We will enter a **DATA LIST** command, followed by a **VARIABLE LABELS** command. These commands will be followed by two others, a **VALUE LABELS** command, and then a **MISSING VALUES** command.

Information about each variable in the dataset should be readily available in the codebook which will have been devised before the data were coded into number form. This information will help us to write the series of commands which will form the data definition file. The codebook for the Gender Attitudes Survey – which provides a working example throughout this book – is reproduced as *Appendix D*). A codebook usually contains the variable names (in the order that they occur within the dataset), the number(s) of the column(s) in which each variable occurs, the missing value assigned to each variable, and any variable- and value-labels we wish to supply at this stage.

The variables must be *defined* (declared, or specified) on a **DATA LIST** command *before* they are labelled, etc. The other commands listed above can be entered in any order and there can be multiple **MISSING VALUE**, **VARIABLE LABELS** and **VALUE LABELS** commands within a single file.

5.1 The DATA LIST command

For data to be usable by SPSS/PC+ they must be defined. Each variable (there can be up to 500) has to be named and the program has to be informed (in a formatting instruction) where to find the relevant dataset and which column(s) of that dataset each variable occupies. Although the data can be included within the data definition file itself (accompanied by additional commands), it is preferable to keep the data in a separate data file and to specify this data file (using '**File =**') on the **DATA LIST** command. This is the method we will use here, and it means that data from the specified data file will be read automatically whenever the **DATA LIST** command is executed.

The **DATA LIST** command will therefore specify the data file in which the data are to be found, the sequence of **variable names** relating to the data, and **formatting instructions** specifying which column(s) within the data matrix each variable occupies. For example:

```
DATA LIST FILE = "study.dat" /ID 1-2 age 3-4 avgrade 5.
```

Thus each variable is named (*id*, *age*, *avgrade*) in the order in which the relevant data appear in the dataset, and after each name the column(s) occupied by the variable are fully specified. In the example above, data for the variable *id* occupy columns 1 to 2 in the data matrix, and the variable *age* is coded in columns 3 to 4. *Avgrade* (e.g. the persons average grade) is represented on just a single column, column 5. For the **DATA LIST**, and for all other SPSS/PC+ commands, no distinction is made between upper-case and lower-case characters. Every command ends with a period (.).

SPSS/PC+ is generally tolerant of extra spaces added into commands as long as they are *between* (rather than within) elements. Also, there must be at least one space between the variable name and the column number (otherwise the column number would be interpreted as part of the variable name).

Variable names

Variable names consist of up to 8 characters (letters or numbers) starting with a letter. Names can be individual specifications (like *sex* and *age*), but numbered series are often included (e.g. var001, var002, var003, etc.). Thus when such series are specified the data list might include, among many other variable names, *item14*, *v1*, *var65* (for 'item 4', 'variable 1' and 'variable 65').

On the data list (and on most other commands) variables can be listed using the **TO** convention. Thus:

`VAR001 TO VAR008`	will create 8 variables.
`V1 TO V105`	will create 105 variables.

But *v1* is treated as a **different** name to *v001*. Thus, after creating a variable *v1* or *v001* (etc.), the exact original form must be used when referring to that variable in any later command.

Formatting instructions

The instruction specifying the column(s) in which a particular variable is to be found is referred to as a **formatting instruction**. In the example:

```
DATA LIST FILE = "trial.dat" / ID 1-4 Sex 5 Age 6-7.
```

we can see that columns are specified by a single number or by a number range (e.g. 1–4). There must be at least one space between the name and the column number(s) and between the column's number(s) and the next variable name.

Where the **TO** convention is used, the columns occupied by the group of variables are assumed to be equally divided between the variables. Thus the specification:

```
VAR001 TO VAR031 10-71
```

indicates that the 31 variables occupy *2 columns each* (notice that the variable numbers and column numbers are inclusive – 'VAR004 TO VAR014' specifies 11 variables; a column specification '35–46' includes 12 columns).

The following example provides the formatting information for 10 '1-column' (or '1-digit') variables, followed by one 3-column variable, followed by *four* 2-column variables:

```
VAR001 TO VAR010 10-19 VAR011 20-22 VAR012 TO VAR015 23-30
```

If data for a single case occupy more than 80 columns, some of the data for that case will have been entered on an additional line (or perhaps over several additional lines). The data list must specify where, in the list of variables, new data lines begin. Column numbering for a second (or third) line then begins from 1 again (*not* 81 or 161). A new line is indicated by a slash (/) and the system then interprets the next 1 as referring to the first column of the second (or third, etc.) line.

Remember that the data for a single variable (for each case) must be *wholly* placed on one line – a single datum (like 34 for age) cannot begin on one line and end on the next. You must specify both the variable name *and* the column allocation on one side or other of the slash.

Thus the following specification is NOT permitted:

```
ID 1-4 VAR001 TO VAR080 5-80 / 1-4 Age 5-6
```

The legitimate form of this expression is:

```
ID 1-4 VAR001 TO VAR076 5-80 / VAR077 TO VAR080 1-4 Age 5-6
```

This tells the system that after reading data relating to *id* and to *var001* to *var076* from the first line, it should read data for *var77* to *var80*, and for variable *age* **for the same subject**, from columns 1–6 of the next line.

Following all of these rules, a valid DATA LIST command might look like this (see also the example in *Appendix F*):

```
DATA LIST File = "survey1.dat"  /  ID 1-4 Sex 5 Age 6-7 VAR001 TO
: VAR006 8-13 VAR007 14-16  VAR008 TO VAR039 17-80 / VAR040
: TO VAR047 1-16 Q1 17 Q2 18-19  Q3 TO Q12 20-29.
```

Here we can see that the **DATA LIST** itself may run over 80 columns. At a convenient point (*not*, for example, in the middle of a variable name or column specification), you may press <RETURN> and continue entering the information on the next line. Additional blank lines are not permitted *within* a command (although they may be inserted *between* different commands in the file). The **DATA LIST** command must end, like other commands, with a period (.).

You should now be able to enter your own **DATA LIST** command. Correct any typing mistakes as you go. The **REVIEW** editing commands are the same as those used when we were entering data to create the data file (press **<F1><r>** for a display).

5.2. The VARIABLE LABELS command

Extended descriptions (labels) can now be provided for some or all of the variables. The labelling of variables (and also of values) is optional and the presence or absence of labels does not affect statistical analyses or any other data manipulation by SPSS/PC+. If they have been provided, however, labels are used to annotate displays of results, including tables, and the SPSS/PC+ output is more readily understood.

Variable labels can be up to 60 characters in length and can include spaces and any other characters. It is best, however to keep them as succinct as possible, especially since the system sometimes prints long labels in truncated (shortened) form. Otherwise, the system prints the labels exactly as they have been specified within the quotation marks. This means that any spelling errors, and any use of upper-case and lower-case letters in the original label specification, will be precisely reflected in the output.

A single **VARIABLE LABELS** command is sufficient to supply labels for all of the variables represented in a dataset, although it is permissible to use several **VARIABLE LABELS** commands within a data definition file. Labels are entered by specifying the name of the variable and then, *within* *quotation marks*, the label to be given to that variable. The variables listed on a **VARIABLE LABELS** command can be in any order and the list therefore does *not* need to follow the sequence provided on the initial **DATA LIST** command (i.e. the order of the variables as they are represented in the data matrix).

Consecutive specifications are separated by a slash (/). For example:

```
VARIABLE LABELS ID "Identification" / SEX "Sex of subject"
/ Age    "Age of subject"
/ VAR001 "Attitude to women's work"
/ VAR002 "Attitude to women's education"
/ VAR003 "Attitude to child-rearing"
etc.
```

Although two or more variable labels can be specified on a single line it is often preferable to allocate a separate line to each variable label, as shown above. A label must not begin on one line and finish on the next. Additional blank lines are *not* permitted *within* the command, and the final variable label specification should end with a period (.).

In some cases it may be appropriate to provide exactly the same label for more than one variable. In this case, rather than writing the same label a number of times, we could enter a list of variable names and supply the 'common' label just once. Thus:

```
/ VAR001 VAR003 VAR004 VAR016 Q5 Q7 "Employment attitudes"
```

Here the six variables named would all taken the common label "Employment attitudes". Such lists can also be used within many other commands (for example when instructing the computer to carry out the same statistical analysis for several variables).

When appropriate, the **TO** convention may be used. Thus:

```
/ VAR001 TO VAR016 "Attitudes to women"
```

Here 16 variables would each take the label "Attitudes to women".

The **TO** convention can be used to specify any group of successive variables in the dataset whether or not they have the same type of name. Thus 'VAR1 TO TIME6' would be valid if it was used to refer to a list of successive variables. The **TO** convention is very useful when a number of successive variables are all to be treated in the same way (in this case all given the same variable label). Thus:

```
/ VAR004 TO Q5 "Attitudes to men"
```

would specify that all of the variables listed on the original data list between *var004* and (up to and including) *q5* should be given the common variable label "Attitudes to men".

Remember that variable labels are optional. You can label as many or as few variables as you wish.

5.3 The VALUE LABELS command

Output which includes only the coded values for a variable – 1, 2, 3, etc. – can be inconvenient to use, for the reader will frequently need to be reminded of what particular values represent. The display or printout (including the output from statistical analyses, and tables and graphs) will be more meaningful when labels have been provided for the various values.

Value labels are optional and their presence or absence does not affect any data analysis. If the data for a variable are truly numbers (as for *age* recorded in terms of years), rather than codes, the **variable label** will generally be sufficient to remind the user of the nature of particular values. In such circumstances value labels would serve no useful purpose. The labelling of values is particularly useful, however, when data have been coded according to a category system. Value labels are specified on a **VALUE LABELS** command and a single command is sufficient to provide labels for many variables (although it is also permitted to use several **VALUE LABELS** commands for different variables in the dataset). The order of variables listed on a **VALUE LABELS** command does not have to follow that on the **DATA LIST** command.

When several variables need to be given the same value labels, you may provide a single set of value label specifications for a list of these variables. In listing these variables the **TO** convention is permitted. Thus:

```
LABELS Sex 1 "Male" 2 "Female" /VAR001 1 "Agree" 7 "Disagree"
 / VAR002 TO VAR005  1 "Positive"   4 "Neither"  7 "Negative"
 / VAR006 VAR008 VAR012 1 "Very much" 2 "A little" 3 "Not at all"
etc.
```

Here the variable *sex* has been given the value label "Male" for the value 1 and the value label "Female" for the value 2. *Var001* has been provided with value labels for just two of its values – 1 (labelled "Agree") and 7 (labelled "Disagree"). The next four variables (*var002* to *var005*) have each been given the same set of value labels for three values (1, 4 and 7) and three further variables have also been provided with labels.

Each set of labels applying to a single variable or to a list of variables must be separated by a slash. Although a *set* of value labels can be split over two or more lines no single label can be split – thus any one *label* must begin and end on the same line. Blank lines must not be inserted within a **VALUE**

LABELS command, and the final value label specification must end with a period (.).

Notice, in the example above, that not all the values associated with a particular variable need to be labelled. Thus for a 1–7 scale of agreement or disagreement, for example, labels may be provided only for the extremes:

```
/ VAR002 TO VAR005  1 "Strongly agree"  7 "Strongly disagree"
```

5.4 The MISSING VALUES command

A missing value is a code assigned to data which are missing from the dataset (for example, where a respondent has failed to answer a questionnaire item). It is a number which is outside the range of valid values of data for that variable. Following the issuing of a **MISSING VALUES** command, SPSS/PC+ will recognize that data coded with this value are missing and will not include the value in, for example, the calculation of an average for the variable. To inform the system that a particular value for the variable signifies missing data, a **MISSING VALUES** command is included within the data definition file. This command lists variables and specifies for each variable listed the appropriate missing value code in brackets:

```
MISSING VALUES Sex (9) / Age (99) / VAR001 TO VAR062 (0)
/ VAR073 TO VAR080 (9).
```

This command instructs the system that 9 is the missing code for *sex* and also for the variables *var073* TO *var080*, that 0 is the missing code for *var001* TO *var062*, and that 99 is the missing code for *age*. No missing value has been provided for some variables which may be presumed to be in the dataset (for example, *var063* to *var072*). The lists of variable names on the **MISSING VALUES** command are generally grouped together in terms of common missing value specifications (9, for example, or 99) and it follows from this that the listing of variables does not have to follow the sequence provided on the initial **DATA LIST** command.

Although it is not mandatory to provide missing values for all variables (or indeed for any variable) it is certainly good practice to do so. In most studies it can be assumed that there will be some missing data.

As well as lists (including the **TO** convention) the keyword **ALL** can be used. Thus:

```
MISSING VALUES All (0).
```

This command would assign 0 as the missing value for each variable in the data list.

5.5 Writing the data definition file to disk

A data definition file is saved ('written to the disk') in the same way as a data file, by using the **REVIEW** command **<F9><w>**. The system will offer the file name originally provided, and if you wish to keep this name for the edited file then you simply need to press <RETURN>. If you have decided to change the name of the file, enter another suitable name, followed by <RETURN>. The data definition file (e.g. "gender.def") will now be stored on the hard disk.

Wc can now exit from **REVIEW** by pressing **<Alt-F10>**, whereupon we will be presented with the SPSS/PC: prompt.

5.6 Testing data and data definition files for errors

When the data definition file has been written to the disk, it is a good idea to test it immediately to see whether it is acceptable to SPSS/PC+. In the early stages of using SPSS/PC+ there are likely to be one or two errors (missing periods – **.** – for example, or misspelt variable names on the labelling commands).

It will be useful, if SPSS/PC+ diagnoses an error, to refer to a printed copy of the data file and the data definition file and to examine these while reading any SPSS/PC+ error message displayed on the screen.

So before attempting to run the data definition file we have created, let us first obtain reference printouts of both the data file and the data definition file. One way of doing this is to leave the SPSS/PC+ system, returning to **DOS** by entering "finish." in response to the SPSS/PC: prompt. Then issue two **DOS** print commands, one for each file. For example:

```
C> type gender.dat>prn
C> type gender.def>prn
```

Next, get back into the SPSS/PC+ system (by entering "spsspc" in response to the **C>**, **D>** or **E>** prompt).

Following the logo and the opening display, we press **<Alt-F10>** to get to our basic start position, the SPSS/PC: prompt. We are now going to attempt to get the system to read our data definition file. If successful we would then be in a position to start issuing statistical commands, etc. To do this we must 'include' the data definition file using the **INCLUDE** command.

Thus, presented with the SPSS/PC: prompt, you can attempt to **INCLUDE** the data definition file. For example:

```
SPSS/PC: include "gender.def".
```

SPSS/PC+ will display each line of the data definition file as it is processed. Each line is also 'diagnosed' as the file is included, and if a problem is diagnosed an error or warning message will be displayed and a low-pitched **Beep** will be heard.

Errors detected when a data definition file is included

As a data definition file is included, the system makes certain checks to see whether the commands in the file conform to certain rules. For example it checks the **DATA LIST** command to see whether different columns have been assigned to different variables, and whether a set of columns specified for a list of variables is wholly divisible by the number of variables listed (thus 'var001 TO var008 2–24' would produce an error message because 23 columns cannot be assigned to eight variables of equal width).

SPSS/PC+ also checks to see whether variables specified on the **VARIABLE LABELS**, **VALUE LABELS** AND **MISSING VALUES** commands are 'known' to the system (i.e. whether they have been declared on the **DATA LIST** command). Thus if the spelling of a variable name on a **VARIABLE LABELS** specification is different from that on the **DATA LIST** command, this will produce a warning message, identifying the item that has caused a problem and informing us of the nature of the problem: **"UNDEFINED VARIABLE NAME–Check for a misspelled variable name"**.

'Warning messages' do not prevent the file from being 'included' but you should study them carefully and try to correct the problem. On the other hand 'error messages' are more serious. For example, if an incorrect column range is specified on the **DATA LIST** command then the command will not be executed and you will be returned to the SPSS/PC: prompt.

Study any error message or warning in conjunction with both the line displayed on the screen and your printout of the data file and the data definition file. These printouts can be amended with any corrections which will need to be made when you re-edit the file.

SPSS/PC+ will detect only certain types of error. Among the common errors which SPSS/PC+ will diagnose at this stage are:

- a period omitted from the end of a command so that the next line is treated as a continuation of that command
- errors in format specification on the **DATA LIST**
- mis-spellings of variable names, so that a variable which appears in a labelling command has not been declared on the data list
- blank lines inserted *within* a multi-line command
- variable labels or value labels split across two lines

Other types of error will be detected by SPSS/PC+ at later stages, for example as it attempts to read the data, or as specific procedure commands are issued. But at no point will SPSS/PC+ be able to detect:

- numerical errors in data entry
- spelling errors within quotation marks
- labels which have been omitted
- errors in the labelling of values or variables

Sometimes SPSS/PC+ will be 'over-enthusiastic' in spotting errors. Thus if a **DATA LIST** command contains a misspelt variable name, any reference to that variable will be diagnosed as erroneous. A subsequent correction of the **DATA LIST** command may therefore prove sufficient to 'correct' a whole series of dependent 'errors' diagnosed by the system.

When errors have been diagnosed by SPSS/PC+ you should **REVIEW** the data definition file, edit out the errors and save the file. To re-edit a file first 'call up' the file with **REVIEW**, citing the name of the file to be edited. For example:

```
SPSS/PC: review "gender.def".
```

After the necessary changes have been made (by moving the cursor to the appropriate places and using the normal **REVIEW** editing functions) you can save the file with **<F9><w>**. The original file name will be presented, and if you wish to retain this name for your amended file you simply need to accept the name that is being offered by pressing <RETURN>. Once an appropriate name has been established for a data definition file, it makes sense to retain it for corrected versions. This will mean that the new version will overwrite the original, preventing you from mistakenly 'including' an old version of the file.

When the corrected file has been saved (using **<F9><w>**), and an exit made from **REVIEW** (by pressing **<Alt-F10>**), the SPSS/PC: prompt will re-appear. Try once again to **INCLUDE** the file. If it now proves acceptable, no error or warning messages will be encountered and you will be presented with another SPSS/PC: prompt. You can now be confident that the file is free of fatal syntax errors although it may retain, for example, spelling errors in labels, or value labels which do not match the values as they were originally coded.

Errors detected when creating an active file

When a **procedure command** is issued (this is usually a command requesting a statistical analysis) the relevant data will be read from the file specified on the **DATA LIST** command and an **active file** will be created by the system. An active file includes the data and the various extra items of information (gathered from the data definition file) which make the data 'understandable' to the system.

One simple procedure command, **LIST**, is discussed later in this chapter. For now, let us use this command to attempt to create an active file. Assuming that *id* is one variable in the dataset you are working with, enter:

```
SPSS/PC: LIST id.
```

This will cause the system to read *all* of the data in the dataset. The message: **"The raw data or transformation pass is proceeding"** is displayed, and you will notice that a number displayed in the top right-hand corner of the screen changes rapidly, increasing by an increment of eight (cases) each time. When the message **"MORE"** is displayed, press any key to release the next screen of information.

You will be presented with a list of all of the *id* values in your data matrix. However, it is possible that as SPSS/PC+ attempts to read the data it will detect certain errors in the data matrix. The system will not detect numerical errors in the dataset, but certain errors can be diagnosed at this stage. The most common of these is the inclusion of one or more inappropriate characters (such as letter, where a number is expected, or an illegal symbol such as @ or %). In this situation, the invalid symbol will be displayed and you will be given the warning: **"INVALID DIGIT READ WITH F FORMAT – Check your data"**. The procedure will not be aborted however, and if you list the values for a variable which has been mis-coded in this way you will see that SPSS/PC+ has automatically assigned a *system missing* value (indicated by a period – **.**). When such a warning is issued you should re-edit the file and correct any mistakes before once again attempting to make the file 'active'.

Another error that may be diagnosed at this stage is reported as: **"NOT ENOUGH RECORDS FOUND TO BUILD A CASE"**. This means that, following the **DATA LIST** command, SPSS/PC+ has been expecting data to be represented within the data file as a complete set of records for each case (a record is a line of data; remember that the data for a single case may extend over one, two, three or more records). The message informs us, however, that an incomplete set of records has been detected. Thus if the data for a case extend over four records, the total number of lines within the dataset is expected to be some multiple of 4. Thus 40, 80, or 88 lines would be accepted by the system, but if 39, 83 or 100 lines are present then SPSS/PC+ will detect that something is amiss and the system will issue the above warning. Again, the procedure is not aborted, but the incomplete case is not included in the active file.

The warning may indicate that you have omitted a line of data, or duplicated one, but the most likely explanation is that you have included one or more blank lines at the end of your data file, and it is worth checking out this possibility first (SPSS/PC+ treats a blank line within the data file as a line of data). To do this, enter **REVIEW** (with the appropriate data file name) and observe whether the cursor, which is automatically placed at the end of the file, is on a blank line below the last line of data. If this is the case delete the extra blank line(s) one at a time using **<F4><d>**, making sure that you do *not* delete the bottom line of data by pressing these keys again when the cursor is *on* the final line of data.

5.7 Obtaining information about the file variables and values

Once an active file has been created you may work with it interactively by responding to the SPSS/PC+ prompt with a variety of commands. Some of the commands available are **procedure commands** and cause data to be processed and analyzed. Other commands merely display information about the variables, and some (like **SHOW** and **SET**, both of which were discussed in Chapter 3) are **operation commands** which provide information about the operating characteristics of the system and allow these to be changed.

Two further commands will be described here. **DISPLAY** is an **operation command** which provides information about any variables which have been 'declared' to the system. The **LIST** command leads the system to display data values for one or more variables. Because the **LIST** command causes the system to process data, it is a **procedure command**.

5.8 The DISPLAY command

The **DISPLAY** command produces a display of information about some or all of the variables included in the active file. In its simplest form, the command produces a list of all the variables in the current active file, together with any variable labels that have been supplied for them. Thus:

```
SPSS/PC: DISPLAY.
```

would produce the following kind of display (and printout, if the printer has been **SET** to *on*):

```
ID          -          * No label *
SEX         -          * No label *
AGE         -          * No label *
FAC         -          Home Faculty
VAR001      -          Women equal to men
VAR002      -          New legislation
VAR003      -          Discrim. as excuse
VAR004      -          Men superior
```

If you wish to interrupt the operation of **DISPLAY** before it has run its full course, press **<^C>** (i.e. the 'Control' key held down and c briefly pressed). This is a useful general 'break' command and will interrupt many types of SPSS/PC+ operation, returning you to the SPSS/PC: command prompt.

If the printer option was *off* during the **DISPLAY** operation, you might wish to repeat the operation after issuing the **SET** command – "set printer on". An alternative way of obtaining a printed copy of the **DISPLAY** output is to request a direct 'Screen print' – a printed copy of the entire screen display. Search for a key which has the label "Scr Pnt" (or something similar). If this key is on the number/function pad of the keyboard, make sure that the number/function pad is switched to the 'function' mode (by pressing the 'Num Lock' key once if necessary). Now press **<Scr Pnt>** once. Provided that the printer is connected to a power source and has been switched on mechanically you should now obtain a printed version of the screen display. You will have to repeat this operation each time you release a new 'page' of screen display (in response to the **MORE** message).

If a variable list, or the keyword **ALL**, is included in the **DISPLAY** command, then fuller information is provided for the specified variables (or for **ALL** of the variables). Thus the command:

```
SPSS/PC: DISPLAY sex fac.
```

might produce the following display:

```
Variable: SEX               Label:  * No label *
  Value labels follow    Type: Number   Width   1  Dec: 0  Missing 9.00
       1.00    male                                2.00    female

Variable: FAC               Label:  Home faculty
  Value labels follow    Type: Number   Width   1  Dec: 0  Missing 9.00
       1.00    arts                                2.00    science
       3.00    social studies                      4.00    other
```

The nature of this **DISPLAY** output will be explained, focussing on the information provided for just one of the variables, *sex*.

The variable name *sex* is provided and we are informed that no label has yet been supplied for this variable. The display also suggests that *sex* is a number variable. This calls for a little explanation. SPSS/PC+ assumes that all data will be in number form unless, for one or more variables, it is informed that data may be in the form of 'alphanumeric characters' (numbers and letters). This use of this option will be addressed in Chapter 16, but for the greater part of this book it is assumed that you will code all of your data in number form. However, the fact that the variable is coded using numbers cannot be taken to imply that the data is truly numerical – suitable for adding, subtracting, etc. In the case of the variable *sex* we know that the numbers are in fact merely codes for 'male' and 'female' and have no valid numerical properties.

The above display illustrates the point that when dealing with data of the 'number' type, the SPSS/PC+ system does not distinguish between numerical and categorical variables. It is up to the user to consider the true nature of the data when requesting analyses and interpreting results. Reporting that the 'average' value of the variable *sex* is 1.473, for example, would *not* be sensible. It cannot be emphasized too strongly that SPSS/PC+ will perform statistical analyses which are invalid and nonsensical if it is requested to do so. It is the user's responsibility to make sure that only reasonable requests are made to the system and that output is interpreted properly.

The 'Dec:' item of the above display reports the number of decimal places assumed to be to the right of the decimal point. In this case this is not a relevant issue (all data are in 'whole number' or 'integer' form, but the system nevertheless reports that there are '0 places to the right of the decimal point'. (Discussion of ways of dealing with non-integer numbers is also left until Chapter 16.) Of more relevance to us at this point is the information in the **DISPLAY** output showing us that the missing value for *sex* has been declared as 9. The display also reports that the value 1 has been labelled "male", and the value 2 "female".

The **DISPLAY** command can be issued at the SPSS/PC: prompt whenever a data definition file has been included, and will produce a useful summary of the variables and labels of which the system has knowledge.

5.9 The LIST command

LIST is a **procedure command** which processes data (it also causes an active file to be created if it is the first procedure command issued after a file has been included). The **LIST** command leads the system to display the data **values** for some or all of the **variables** and for some or all of the **cases** in the file. In its simplest form (the single command word **LIST.**), the **LIST** command produces a list of values for every variable and for every case in the active file. A **LIST** command can also be used to display the data relating to *specified* variables. Thus (using the Gender Attitudes files):

```
SPSS/PC: LIST id sex var001 prs04.
```

produced the following display:

```
        ID SEX VAR001  PRS04

         1   1      2      2
         2   2      1      2
         3   1      5      2
         4   2      2      2
         5   2      2      1
 ...
        50   1      2      2
```

If a list of values is required for only *some* of the *cases*, the **LIST** command should specify those cases for which the data are required, using a **Cases** subcommand. The system assumes that listing of cases should begin at the first case in the file and end at the last unless it is instructed otherwise. Thus:

```
SPSS/PC: LIST CASES= TO 5.
```

or

```
SPSS/PC: LIST CASES= from 1 TO 5.
```

or

```
SPSS/PC: LIST CASES= 5.
```

would all produce a list of the values of *all variables* (no variable names have been specified) for the first five cases in the file. However:

```
SPSS/PC: LIST CASES= FROM 5.
```

would list the values for all cases in the file, except the first four cases, and:

```
SPSS/PC: LIST CASES= FROM 5 TO 15.
```

would list the values for the 11 cases specified.

If both a variable list *and* a **Cases** subcommand are included within a single **LIST** command they must be separated by a slash ('/'). Thus:

```
SPSS/PC: LIST sex fac var001 to var010/ cases= from 10 to 20.
```

This would produce a list of values for each of the 12 variables specified for the 11 cases specified.

5.10 Using active files for analysis

When a data definition file has been successfully **INCLUDED** and the first procedure command is issued, the data are read in order to create an active file. Any further procedures that are requested then use the data from the existing active file (usually to perform a statistical analysis). One simple procedure, **FREQUENCIES**, produces tables of frequencies and percentages for the various values of a variable. A **FREQUENCIES** analysis of the data for the single variable *sex*, for example, could be obtained by issuing the following command:

```
SPSS/PC: frequencies sex.
```

This would produce a display of the following type:

```
SEX
                                                  Valid      Cum
   Value Label        Value  Frequency   Percent Percent   Percent

male                      1         74      54.0     54.0     54.0
female                    2         63      46.0    100.0    100.0
                             - - - - -     - - -    - - -
                      TOTAL        137     100.0    100.0
```

It can be seen that the **FREQUENCIES** output provides value labels and the frequencies of valid (i.e. not missing) cases coded with each value. These frequencies are then expressed in terms of a percentage of *all* cases and a percentage of valid cases only. Valid percentages are also added sequentially to produce the cumulative percentage. As we will see in the next chapter, the above example represents the basic or default output from the **FREQUENCIES** procedure. Fuller output (including additional statistics, bar-charts, etc.) can also be requested.

If you have created a data file and a data definition file relating to your own study you may wish, at this stage, to obtain frequencies for several more variables. If you have not yet created such files, this would be a good point at which to do so. You will then be able to analyze your data as you read through the remaining chapters of this book. If, as yet, you have no data of your own you may wish to use **REVIEW** to produce your own copy of the data file and data definition file from the Gender Attitudes Survey (these are reproduced in *Appendix E* and *Appendix F*). When you have written your files they should be saved in your own directory on the hard disk. The data definition file can then be **INCLUDED** in any future SPSS/PC+ session by issuing a command such as:

```
SPSS/PC: include "gender.def".
```

When this file has been included, and a procedure command has been issued, you will have created an active file that can be used for analysis and manipulation. In the next chapter we will begin to explore in detail some of the procedures which can be used to provide statistical analyses of the dataset.

6 Statistical Analysis I: FREQUENCIES and CROSSTABS

In the previous two chapters we have been concerned with writing the data file and the data definition file, using **REVIEW**. When a data definition file is included (by issuing an **INCLUDE** command, specifying the file name, at the SPSS/PC: prompt) SPSS/PC+ reads the file, checking each line. If the file has been successfully included, then issuing any **procedure command** (such as **LIST**, or a statistical command) causes the data to be read and creates an **active file**. The active file contains the data in a form which is meaningful to the system. The active file includes the values corresponding to each variable for each case (including the missing values) and any variable- and value-labels associated with particular variables. Data within an active file can be manipulated and analyzed by issuing the appropriate SPSS/PC+ commands interactively i.e. by keying them in response to the SPSS/PC: prompt.

Although there are many more things to learn about file creation and data manipulation, this is a good point at which to begin to consider how SPSS/PC+ can be used to perform statistical analyses.

Some readers will be very familiar with statistics, whereas others will have only a basic knowledge. For serious work it is important that you do not use SPSS/PC+ beyond your own statistical capabilities. But playing with data, particularly by using the power and convenience of SPSS/PC+, is a good way of whetting your appetite to know more about statistics. There are many good statistics textbooks, including the *SPSS^X Introductory Statistics Guide*, and the *SPSS/PC+ Manual* provides much useful information, particularly about the more complex forms of analysis.

If when using this book you meet any term with which you are unfamiliar (for example, in this chapter we use 'chi-square', 'significance', 'degrees of freedom', 'parameters', 'crosstabulation', 'variance', etc.) consult the Glossary (*Appendix G*) for an explanation.

At this stage you will probably not be content to analyze 'dummy data'. The analysis of a series of random numbers provides little interest. If, therefore, the data definition file you have created is of this type you might prefer to create another file, this time containing and defining real data. You can collect data in a study of your own or you can use that from the Gender Attitudes Survey. The complete dataset obtained from this survey is provided in *Appendix E*, and the data definition file is provided in *Appendix F*. The examples given in this and many later chapters are based on this survey.

6.1 Using procedures

After creating a suitable data definition file, the first step towards using the file for analysis is to include the file with the **INCLUDE** command. For example:

```
SPSS/PC: include "gender.def".
```

When the SPSS/PC: prompt returns, the SPSS/PC+ system is awaiting a further command. The first **procedure command** that is issued (such commands include **LIST**, which was discussed in the previous chapter, and the various statistical commands that we meet in this and later chapters) first causes the data to be read (automatically, from the data file) so that an 'active file' is created. The system then carries out the instruction specified on the procedure command (e.g. it lists values, or performs statistical tests). The active file is not a file stored on the hard disk but remains 'active' in the computer memory (the original copy of the data, in the data file, remains intact).

Once the active file is in the computer memory, further procedure commands cause data to be analysed, tabulated, and transformed in accordance with the user's instructions. In this chapter we will consider the use of two procedure commands – **FREQUENCIES** and **CROSSTABS**.

Each procedure command runs a set part of the SPSS/PC+ program, often subjecting the data to some type of statistical analysis. Most procedure commands can be supplemented with one or more subcommands which allow users to tailor the procedure to their own needs. Two of the most useful, and generally used, subcommands are **Options** and **Statistics**. The **Options** subcommand is frequently used to specify how missing data should be dealt with and to request a particular form for the output display or printout. The **Statistics** subcommand allows the user to request statistical information additional to that provided by the procedure 'by default'.

We will start to use procedure commands (including subcommands) interactively with the SPSS/PC+ system (i.e. in response to the SPSS/PC: prompt). Later (in Chapter 9) we will see that procedure commands and subcommands can also be written into a separate file or added directly into data definition files (such files are referred to as 'command files'). At this stage, however, we will assume that the procedure commands and subcommands are to be issued interactively.

After issuing a complete command, ending with a terminator period (**.**), a number will appear in the top right hand corner and will change rapidly, indicating that cases are being processed and analyzed. When this number reaches the total number of cases in the file, and after a short pause during which further processing is taking place, the first 'page' of results will be displayed on the screen. If the printer is set to *on* then the first page of results will also be printed.

After the first 'page' has been displayed, the operation will be halted. The message **MORE** will be presented at the top right hand corner of the screen and a high-pitched beep will sound. Pressing any key on the keyboard will release the next 'page' of information.

If the user wishes to examine the results of the procedure when they are later printed from a file, rather than as successive screen displays, then the pause between 'pages' (accompanied by the **MORE** message) will be redundant. The pause facility can be switched off (so that all of the output is displayed without a pause) by using the **SET** command (*before* issuing the procedure command) to 'set more off.':

```
SPSS/PC: set more off.
```

Errors detected when procedure commands are issued

We have already considered, in the last chapter, errors that may be detected by the system when you attempt to include the data definition file and when you attempt to create an active file.

When a procedure command is issued, further errors may be detected by the system. Procedure commands may instruct the system to perform analyses on a large number of specified variables, and the command may contain several rather complex subcommands. A single procedure command may thus extend over several lines, and for each command and subcommand there are particular rules of 'syntax' which specify the general form that the instruction to the system must take.

Thus there is lots of room for error, and it is not uncommon for complex procedure commands to be entered more than once before the user gets it right. Common errors include:

- one or more of the key words in the command is misspelt, so that the command is not recognized
- one or more of the variable names is mis-spelt (so that the a variable specified in the command is held not to exist)
- the requested procedure is not included among those available on your computer
- the command exceeds certain system limitations. Thus for many procedures there are limits to how many variables can be cited in a single command (if you make this error, the answer is to call for the analysis of fewer variables by issuing a series of similar commands, each citing just

some of the variables quoted in the original command).

Fortunately, when the system diagnoses an error of any of the above kinds, it issues a message informing the user of the nature of the problem. It is then a simple matter to re-enter the command in such a way that the problem does not recur.

We will now consider the use of two SPSS/PC+ procedures for statistical analysis – **FREQUENCIES** and **CROSSTABS**.

6.2 The FREQUENCIES procedure

It is best to begin the analysis of a dataset by examining the basic features of single variables. For coded variables we might wish to know the frequency distribution over the range of different values; this will allow us to examine the composition of the sample in terms of sex, age and other such parameters and to find out how many cases have any particular value for a variable. We might also require basic statistical information such as the mean (= average) and standard deviation of individual variables. The **FREQUENCIES** procedure produces tables of frequency counts for the values of any variable(s) specified and can also be used to display a variety of summary statistics for such variables.

To use **FREQUENCIES** we first need to **INCLUDE** the relevant file. For example:

```
SPSS/PC: include "gender.def".
```

The SPSS/PC+ prompt will again be presented and we can respond with the procedure command. Thus:

```
SPSS/PC: frequencies sex.
```

When <RETURN> is pressed there will be a pause while SPSS/PC+ processes the data and the following results table will then be displayed:

```
SEX      Subject sex

                                                      Valid       Cum
 Value Label            Value  Frequency   Percent  Percent   Percent

Male                        1         22      44.0     44.0      44.0

Female                      2         28      56.0     56.0     100.0
                               --------   -------  -------
                        TOTAL         50     100.0    100.0

Valid Cases          50    Missing Cases    0
```

We can see that the total population of valid cases is 50 and that 22 of these are male and 28 are female. The table also includes the frequency of each *sex* group as a percentage of the total population (valid + missing) and of the total valid (only) population. The cumulative percentage is also displayed.

When the output from **FREQUENCIES** (or another command) is displayed, you may find that the total number of cases (valid cases + missing case) reported slightly exceeds the number of cases actually represented in the data file. This probably means that you have one or more 'extra' blank lines in your data file, which the system is mistaking for data. In this case re-edit the data file to delete the extra blank lines (start with the cursor at the very end of the file – a blank line – and use **<F4><d>** to delete each blank line, making sure that you stop when the cursor is *on* the bottom data line).

Several variables can be listed within a single **FREQUENCIES** command (and the **TO** keyword can be used). Thus the following list produces a frequency table for each of the eight variables listed:

```
SPSS/PC: frequencies sex  var001 to var006 st02.
```

Statistical information can be obtained by issuing a **Statistics** subcommand. This specifies which statistics are to be displayed by quoting one or more keywords (for example 'stddev' to obtain the standard deviation).

The **Statistics** subcommand is specified after the main command following a slash (/) and takes the form:

```
: / statistics = median stddev minimum.
```

If this is the final subcommand to be issued then it should end with the terminator period (as above).

Thus the complete command would be:

```
SPSS/PC: frequencies sex  var001 to var006 st02 <RETURN>
```

(there will be a pause while SPSS/PC+ does some initial processing) and then:

```
: / statistics = median stddev minimum.
```

There are many ways of finding out which **Statistics** are available for any procedure. Lists of the available **Statistics** are provided in the *SPSS/PC+ Manual* and a selected list of the more frequently used **Statistics** (and **Options**, where they are available) is also provided at the end of this book (*Appendix J*).

In addition to these 'off-line' information sources, 'on-line help' can be obtained by using the SPSS/PC+ HELP facility that is available from within **REVIEW**. However, this is most useful when we are issuing commands not interactionally but in a **command file**. The use of command files is introduced in Chapter 9, and we will then consider the use of the on-line help facility.

The **Statistics** available with **FREQUENCIES** include 'measures of central tendency' (mean, mode and median), 'measures of dispersion' (standard deviation, variance), and measures of the shape of the frequency distribution (skewness and kurtosis). For a general explanation of such statistical terms see the Glossary – *Appendix G*. The **FREQUENCIES Statistics** are also briefly described in Table 6.1. The 'keyword' item in the Table refers to the word used to request the statistic within the **Statistics** subcommand.

In specifying which statistics are to be performed on the data, the keywords **ALL** and **NONE** can be used. If a **Statistics** subcommand is issued without specifying any particular statistics then certain default statistics – MEAN, STDDEV, MINIMUM and MAXIMUM – are displayed. If no **Statistics** subcommand is included then no statistics are displayed.

It should be noted that data relating to particular variables will be suitable only for *some* of the analyses available, but that SPSS/PC+ will, at the user's request, perform many totally invalid analyses on data. Thus if a specific request is made, or a relevant default not over-ridden, SPSS/PC+ will happily produce a mean (average) for a categorical variable. Thus, using the data from the Gender Attitudes Survey, a mean of 1.56 was obtained for the variable *sex*. If we had chosen instead to code males as 2 and females as 1 – an equally arbitrary strategy – the mean for the same population would then have been represented as 1.44).

Such figures, however, are clearly quite meaningless. In requesting a mean (or standard deviation, or median, etc.) for such data we have completely ignored the vital difference between *coded* variables (such as *sex*) and genuine *numeric* variables (such as *salary*). The point is that SPSS/PC+ is not aware of how numbers have been used. It will perform the analyses it is asked to perform, as long as certain basic criteria are fulfilled. It has no 'appreciation', in cases such as the example we have given, of whether data are numeric or categorical. The responsibility for the appropriateness of the analyses performed, therefore, rests with the user rather than with the system.

Although there are many clear-cut instances in which particular types of analysis are appropriate or inappropriate, there are situations where opinions on the appropriateness of employing particular procedures differ sharply. Some statisticians, for example, are happy to use a full range of statistical tests on 7 point scale data whereas others consider such data to be categorical rather than numerical and believe that arithmetic operations should not be performed.

There is no **Options** subcommand for the **FREQUENCIES** procedure, but there are various ways of tailoring the output. Thus simple graphical displays

Table 6.1 Frequencies Statistics

Keyword	Name	Description
MEAN	Mean	the arithmetic average.
STDDEV	Standard deviation	a measure of the spread of the frequency distribution.
MODE	Mode	the most frequent value for this variable within the dataset
MINIMUM	Minimum	the lowest value.
MAXIMUM	Maximum	the highest value.
RANGE	Range	the difference between the lowest value and the highest value.
SEMEAN	The standard error of the mean	a measure of the confidence we can have that the average based on this sample is a true representation of the average of a larger theoretical population.
VARIANCE	Variance	another measure – along with STDDEV – of the spread, or variation, within the data.
SKEWNESS	Skewness	a measure of how 'lop-sided' the data are – i.e. a tendency for more data to occur towards the lower or upper end of the range rather than being evenly distributed or balanced equally on either side of the mid-point of the range.
SESKEW	The standard error of the skewness	a measure of the confidence we can have that the skewness of this sample is a true representation of the skewness of a larger theoretical population.
MEDIAN	Median	The value which is a mid-point, dividing the population into 50% below the median and 50% above the median.
KURTOSIS	Kurtosis	The extent to which the data frequencies 'peak' around the middle value or are 'flat' across the range of values.
SEKURT	The standard error of the kurtosis	A measure of the confidence we can have that the kurtosis of this sample is a true representation of the kurtosis of a larger theoretical population.
SUM	Sum	The sum of all the data points for the variable.

of the frequency distribution (in the form of 'barcharts' and 'histograms') can be requested.

For example, to request a simple barchart:

```
SPSS/PC: frequencies fac sex var001 to var006 prs05
       : / barchart.
```

This would produce, as well as the normal display information, a barchart for each of the variables listed on the **FREQUENCIES** command. The raw frequency of each value for that variable is plotted on a horizontal axis, and value labels are included in the output.

The output from the **FREQUENCIES** procedure (including that obtained from the **Barchart** and **Histogram** subcommands) can be tailored in a number of ways. The **SET** command, for example, can be used, *before* the procedure command is issued, to change the symbol used to print a histogram or barchart. Full information is available in the *SPSS/PC+ Manual*.

If two or more subcommands are used within the same command they must be separated by a slash. Thus:

```
SPSS/PC: frequencies fac sex var001 to var006 prs05
       : / statistics = mode range
       : / barchart.
```

Only the final subcommand is followed by the terminator.

6.3 The CROSSTABS procedure

We often wish to know the frequencies of values with respect to two or more coded variables simultaneously. Thus if we have information on the sex of our respondents and on their marital status we might wish to display a table which provides the frequencies of married males, married females, divorced males and divorced females. This would enable us to examine the question of whether within our sample the proportion of males who are divorced is greater than the proportion of females who are divorced. If this proved to be the case then we might conclude that (for our sample) sex and the state of being divorced are not independent, or, to put it another way, that there is an association between sex and divorced status. A crosstabulation – or joint frequencies table – will display such information in a convenient form. Such a crosstabulation can be obtained by using the **CROSSTABS** procedure. In its simplest form it simply names a variable, uses the keyword **BY**,and then names a second variable. Thus:

```
SPSS/PC: crosstabs fac BY sex.
```

A crosstabulation table displays the joint distribution of two (or more) variables each with a limited number of (usually categorical) values. In the above example, because the variable *sex* can take either of two values, and *fac* can take one of three values (in the Gender Attitudes Survey – 'Arts', 'Science' and 'Social Studies'), a crosstabulation of the two variables will be a 2 X 3 table with six cells:

```
                      Male         Female
                 ......................... Row Total
                 :           :            :
          Arts   :     6     :     12     :   18
                 :...........:... ........:
                 :           :            :
       Science   :     8     :      4     :   12
                 :...........:............:
                 :           :            :
  Soc. Studies   :     8     :     12     :   20
                 :...........:............:

       Column         22           28       TOTAL    50
        Total
```

The three rows correspond to the three valid values for *fac* and the two columns correspond to the valid values for *sex*. There are thus six cells. Reading the information from the table we can see that there are 50 subjects, 22 male and 28 female. Of our population 18 are from the arts faculty, 12 are from the science faculty and 20 are from the social studies faculty. From the information within the cells we can see, for example, that there are 12 female arts students in the population, and 8 male science students.

The above table was obtained by using the **CROSSTABS** command:

```
SPSS/PC: crosstabs fac BY sex.
```

Had we specified the alternative order ('*sex* by *fac*') then *sex* would have been displayed as rows and *fac* as columns:

```
SPSS/PC: crosstabs sex BY fac.

   FAC             Arts          Science   Soc. studies
               ....................................... Row Total
 SEX           :          :           :           :
        Males  :    6     :      8    :     8     :   22
               :..........:...........:...........:
               :          :           :           :
      Females  :   12     :      4    :    12     :   28
               :..........:...........:...........:

        Column     18          12          20       TOTAL 50
Total
```

The basic form of a **CROSSTABS** command is:

CROSSTABS variable name (or list) BY variable name (or list)

For example:

```
SPSS/PC: crosstabs prs01 to prs10 BY sex fac.
```

This will produce 20 separate crosstabulations (i.e. each of the 10 variables *prs01* to *prs10* will be crosstabulated with *sex* and, separately, with *fac*).

Several crosstabulations may be requested within the same **CROSSTABS** command, even if none of the variables in different crosstabulations are the same. In this case a slash is used to separate requests for different tables. For example:

```
SPSS/PC: crosstabs fac BY prs01 / sex BY yr.
```

So far we have considered only crosstabulations employing two variables. It is also possible to request crosstabulations which display the frequency distributions according to three or more variables. Thus:

```
SPSS/PC: crosstabs fac BY sex BY st04.
```

would give a single 'three-dimensional' table, in which each cell would include the number of cases for one faculty, one sex and one value of the variable *st04*. Thus with three faculties, the two sexes, and two possible (valid) values for *st04*, there would be 3 X 2 X 2 (= 12) cells.

Up to ten dimensions can be specified in any one command. However, as the number of cells increases the average number of cases per cell will obviously decrease and a low number within one or more cells (and especially, the presence of 'empty cells') will prohibit the legitimate use of many of the statistical tests available for **CROSSTABS**.

If no subcommand is included in the overall procedure command, then the procedure will operate with all of its defaults in effect. By default, the **CROSSTABS** output includes tables with frequency counts (but not percentages) in each cell, marginal (row and column) counts and percentages, variable labels and value labels, and the numbers of valid and missing cases.

Thus the following table presents the complete default output from the commands:

```
SPSS/PC: crosstabs sex BY prs10.
```

```
Crosstabulation:      SEX        Subject sex
                   By PRS10      Trust difficult (SELF)

            Count  |TRUE     |FALSE    |
 PRS10->           |         |         | Row
                   |       1 |       2 | Total
SEX        --------+---------+---------+
                1  |     8   |    14   |    22
  Male             |         |         |  44.0
                   +---------+---------+
                2  |    12   |    16   |    28
  Female           |         |         |  56.0
                   +---------+---------+
            Column      20        30         50
             Total    40.0      60.0      100.0

Number of Missing Observations =          0
```

The **CROSSTABS** command may include a number of subcommands, including a **Statistics** subcommand.

The **Cells** subcommand is used to specify that extra information pertaining to each cell of the crosstabulation should be included in the display. Options here include 'row', 'column', 'total' and 'expected'.

If 'row' is specified on a **Cells** subcommand then the information displayed within each cell will include the number of cases within that cell, displayed as a percentage of all of the cases in that row. Similarly, if 'column' is specified then the cell display will include the number of cases in the cell expressed as a percentage of all cases in that column, and 'total' will lead to the inclusion of the number of cases in the cell as a percentage of the total cases in the table.

If the row variable and the column variable were completely independent, then we would expect the numbers of cases within each cell to be 'evenly' distributed (i.e. for each column, the same proportion of cases occurring within each row, whichever column was examined).

The following fictional table represents such independence.

```
    EYECOL          Brown eyes   Blue eyes
                  ......................... Row Total
  SEX             :            :           :
         Males    :    50      :     25    :    75
                  :............:...........:
                  :            :           :
       Females    :    10      :      5    :    15
                  :............:...........:

       Column          60            30   TOTAL 90
        Total
```

Although, from this table, we can see that there are more 'male' than 'female' cases, and many more 'brown-eyed' people than 'blue-eyed' people, notice that, for both sexes, there are precisely five times as many 'brown-eyed' subjects as 'blue-eyed' subjects. Thus there is no association between the variable *sex* and the variable *eyecol* in this fictional example.

There is another way of demonstrating the independence between these two variables. If, for any cell, we multiply the relevant 'row total' by the relevant 'column total', and then divide by the total for the whole table (90, in this case) we will end up with the number that is actually written in the cell (thus, for the cell with the smallest total – (15 X 30) / 90 = 5). This formula – 'row total X column total divided by grand total' – can be applied to any cell to derive the 'expected frequency' within the cell (the count that would be found in the cell if the row variable and the column variable were totally independent). The difference between this expected frequency and the actual count within the cell ('the observed frequency') is known as the 'residual frequency'.

The **Cells** subcommand can be used to request the inclusion, within the cell display, of the expected value and the residual value ('resid'). Thus a **Cells** subcommand might read:

```
/ cells = row column expected resid.
```

Contrast the display in the table on p. 82 above with the equivalent display obtained when certain **Cells** options are requested. In this case the command:

```
SPSS/PC: crosstabs sex BY prs10
       : / cells = expected row column total.
```

will produce the following output:

```
Crosstabulation:       SEX        Subject sex
                    By PRS10      Trust difficult (SELF)

                Count |
              Exp Val |
              Row Pct |TRUE      |FALSE    |
     PRS10->  Col Pct |          |         | Row
              Tot Pct |        1 |       2 | Total
SEX           --------+----------+---------+
                    1 |        8 |    14   |   22
     Male             |      8.8 |  13.2   |  44.0
                      |     36.4 |  63.6   |
                      |     40.0 |  46.7   |
                      |     16.0 |  28.0   |
                      +----------+---------+
                    2 |       12 |    16   |   28
               Female |     11.2 |  16.8   |  56.0
                      |     42.9 |  57.1   |
                      |     60.0 |  53.3   |
                      |     24.0 |  32.0   |
                      +----------+---------+
               Column         20      30       50
                Total       40.0    60.0    100.0
```

Each cell now contains 5 items of information and a guide to the various items in provided in the top left corner. The first item in each cell is the cell count (as in the default table). The second item is the frequency which would be expected in that cell if the two variables (in this case *sex* and *prs10*) were statistically independent. The remaining items are percentages – the cell frequency is expressed as a percentage of the row total, the column total and the population total.

A number of statistical tests can also be requested on a **Statistics** subcommand by citing appropriate specifications (which, in SPSS/PC version 4, are made by name rather than, as in earlier versions, by reference numbers). These include the 'phi coefficient' (specify **phi**), the 'uncertainty coefficient' (specify **uc**), the 'contingency coefficient' (specify **cc**) and the most widely used crosstabulation statistic 'the chi-square test' (specify **chisq**). Information on the other available statistics will be found in the *SPSS/PC+ 4.0 Base Manual*. We will limit further discussion to a consideration of the use of the chi-square test.

6.4 Inferential statistics: the use of chi-square

Some statistics are merely summaries or descriptions of the data. They may cite an average, for example, or summarize the 'spread' of the data. Such statistics (which include the mean, median, mode, range, standard deviation, kurtosis, etc.) are known as 'descriptive statistics'. Other statistics allow us to draw conclusions about whether, for example, two samples differ, or whether there is an association between two variables which cannot simply be attributed to chance. Such statistics emerge as the result of a statistical test, and they allow us to make an inference. Hence they are called 'inferential statistics'. Chi-square is an example of an inferential statistic and we will meet with many others.

When the value for an inferential statistic has been calculated we usually wish to know whether that value represents something 'significant'. It is in the nature of such statistics that it is possible to compute the probability of any particular value emerging from essentially random data 'by chance'. Generally, the higher the value of such a statistic the less frequently it would emerge by chance. Thus if a high value (of chi-square, for example) is obtained using the data entered into the statistical equation then it is likely that this reflects some real pattern in the dataset rather than mere chance.

A common criterion for deciding whether such there is a pattern in the data involves applying a '1 in 20' rule. If the value obtained for the statistic is higher than we would expect to emerge by chance alone on 5% (1 in 20) or fewer occasions then we infer that our value is probably not the result of chance but is likely to reflect a real effect (or pattern). We say that such a result is significant ('at the 5% level of probability'). We might wish to play safer and adopt a more stringent criterion by using a '1 in 100' rule. This would demand that a higher value of the statistic be obtained (such as would occur by chance on only 1% of occasions – the probability is less than .01) before inferring that a pattern or effect has been demonstrated (values that exceed the relevant threshold are said to be 'significant at the 1% level of probability').

Typically, when inferential statistics have been calculated, the SPSS/PC+ output also provides a value for the probability of the statistic obtained. Thus if we are using a 5% probability criterion (which could be written .05) the probability values .035 and .021 (i.e. less probable than .05) would be sufficient for us to conclude that the pattern or effect was significant. Probability values of .06 or .62, however, would not be sufficiently low (the observed pattern would not be sufficiently improbable) for us to conclude that an effect was present. We would not be able to conclude that there was 'no effect' but would merely have to admit that our data did not show a significant effect.

If the 1% (.01) criterion is adopted, then SPSS/PC+ would have to return a probability value lower than .01 (e.g. .003 or .0019) before we accepted that a significant effect had been demonstrated.

The **Statistics** available for the **CROSSTABS** procedure include a number which assess the degree of association between the variables specified by the command. Thus for chi-square (specified on a **Statistics** subcommand by 'chisq'), a significant result indicates that there is a relationship between the 'row variable' and the 'column variable'. Thus if, in the '*sex* by *fac*' analysis, a chi-square test had been requested and a significant result obtained, this would have indicated that, for the data within our dataset, there was a significant association between respondents' sex and their home faculty. An examination of the crosstabulation might have revealed, for example, that more females tended to be from the arts and social studies faculties and more males from the science faculty, and we would know (because of the low chance probability of the chi-square value obtained) that the observed faculty distribution of the sexes was unlikely to have been a result of mere chance.

An analysis of the following table *fails* to indicate a *significant* association between *sex* and *prs07* ('trouble sleeping'):

```
   SEX      Subject sex
By PRS07    Trouble sleeping (SELF)

                         TRUE      FALSE      Row
                     .....................  Total
                     :         :         :
          Male       :    7    :   15    :   22
                     :.........:.........:
                     :         :         :
        Female       :    6    :   22    :   28
                     :.........:.........:

         Column          13        37
          Total                  TOTAL 50

Chi-square   D.F.   Significance    Min E.F.   Cells with E.F. < 5

    .25667     1        .6124          5.720      None
    .69120     1        .4058        (Before Yates Correction)
```

This output table needs some explanation. We see that two values have been calculated for chi-square. This is because the table is a 2 X 2 crosstabulation. The bottom line includes a normal chi-square calculation (i.e. 'Before Yates Correction'). The penultimate line includes an alternative calculation using a modified formula which, for technical mathematical reasons, incorporates

'Yates' Correction'. When the correction factor has been applied (and it is applied only to 2 X 2 tables) we can see that the effect is to reduce the value of chi-square and hence to increase the probability estimate (to .6124). Increasing the probability in this way makes it *less* likely that a significant result will be obtained, and the use of Yates' correction is therefore a conservative option (making it less likely that we will attribute significance to a data pattern which actually results from the effects of chance). There is some controversy about whether it is necessary to apply Yates' correction or not, but some authorities do insist that it should be employed whenever we perform a chi-square analysis on a 2 X 2 table.

In addition to the two chi-square values, with their associated probabilities, the output display also includes values for D.F. (degrees of freedom – see Glossary) and for the 'expected frequencies' of the cells in the table. The chi-square formula calculates expected frequencies and then compares these with the actual (observed) frequencies in the cells. If the differences between the expected and observed values are high then a high value is obtained for chi-square. The information in the table includes the minimum expected frequency and the number of cells which have an expected frequency of less than 5. If there are cells with a lower expected frequency then some statisticians would doubt the validity of the chi-square statistic obtained.

The chi-square value for the above table (after applying Yates' correction) is 0.257. This corresponds to a probability of .6124 which, being far higher than the .0500 threshold for significance, indicates that – according to our dataset – one sex is not more likely to report sleeping difficulties than the other. This reflects the fact, observable within the crosstabulation, that roughly the same *proportion* of males as females say that they have trouble sleeping.

In the next table, however, we see that there is an apparent association between the two Gender Attitudes Survey variables *sex* and *prs02*:

```
       SEX    Subject sex
   By PRS02   Sex thoughts often (SELF)
                          TRUE      FALSE       Row
                      ....................     Total
                      :         :         :
            Male      :   15    :    7    :    22
                      :.........:.........:
                      :         :         :
          Female      :   10    :   18    :    28
                      :.........:.........:
           Column         25        25
            Total                         TOTAL 50
```

continued

Chi-square	D.F.	Significance	Min E.F.	Cells with E.F. < 5
3.97727	1	.0461	11.000	None
5.19481	1	.0227	(Before Yates Correction)	

The information in this table indicates that a higher proportion of males than females in the sample claimed that they thought about sex often. The chi-square value for the table (with Yates' correction applied) is 3.977 and the probability associated with this value is 0.0461. Thus the probability of such a pattern emerging purely by chance (for example, from sets of random data) is just *less* than the .05 (= 5%) criterion and we might therefore wish to conclude that the data show a significant relationship between the subject's sex and the likelihood of their claiming to think about sex often.

Other **Statistics** available with **CROSSTABS** also measure the degree of association between the crosstabulated variables. They make different assumptions about the data, however, and for some data some of the tests are considered more suitable than others. The choice of test will depend on the needs of the user and on certain conditions within the data.

Thus the legitimate use of the chi-square test demands that no cell is empty and the use of chi-square on a table containing any cell with an expected frequency of less than 5 is controversial. Once again it is worth making the point that SPSS/PC+ will often calculate invalid statistics. Thus it will perform chi-square tests even if certain assumptions of the test are infringed.

Although SPSS/PC+ provides a wide range of tests, there would never be a situation in a real research study in which all of the available statistics were appropriate or necessary. The needs of most users will probably be satisfied by employing only one or two of the statistics available for **CROSSTABS**, and chi-square will almost certainly be found to be one of the more useful.

6.5 Using the listing file to view results

You may now have started to examine and analyze your dataset, perhaps using **FREQUENCIES** and **CROSSTABS**. The question arises of how you will store the output from your procedures so that you may read it, copy it, edit it, and print it after your 'session' with SPSS/PC+ has finished. (A 'session' starts when you enter SPSS/PC+ and finishes when you exit. Thus if for some reason you move back and forth between SPSS/PC+ and **DOS** then you may have several 'sessions' at one sitting.)

There are a number of options. If the printer is set to *on*, for example, then the results of any analysis that are displayed on the screen will also be printed at the same time. You can then take the printed output away with you to examine at a later time.

There is, however, a much better way of doing things. One of the most useful features of SPSS/PC+ is that it automatically creates a **listing file** which contains a complete record of all of the **display output** for the current session, together with any warnings or error messages which have been issued. The pages within the listing file are numbered and the complete listing for a session is available on a file named (by default) "SPSS.LIS".

This is an ASCII file, which means that it is editable by word-processors and editor programs (including **REVIEW**). And the fact that it is editable means that you can annotate it with text, paginate the material in an orderly fashion, and change the order of presentation. If you use a word-processor you can also add typographical features such as **bold** and *italics* (or underline). Furthermore, the file can be copied to a floppy disk and taken to another machine for editing and printing.

Since, by default, the listing file is always given the same name ("SPSS.LIS"), and since a file is deleted (overwritten) whenever a file with the same name is written to the same directory, it follows that the listing file from the last session of SPSS/PC+ (your own session, or someone else's) will be deleted when the next SPSS/PC+ session starts.

To preserve a listing file, therefore, it should be given another name, either at the beginning of a session or afterwards, when you exit to **DOS**. An alternative name for the listing file can be specified at the beginning of the session, using the **SET** command. For example:

```
SPSS/PC : set listing "gender.lis".
```

The use of such **SET** commands prevents the loss of the records at the beginning of the *next* session when an existing file with the default name "SPSS.LIS" is initialized. If an alternative name has been provided, the listing file created during the session will remain on the disk until deliberately deleted or overwritten by another file which has been given the same name. Thus to keep a permanent record of the commands and output from each session, *different* names must be given to the listing file on each occasion.

If the listing name is **SET during** a session then the named files will contain the listing *from that time on*. Records of the earlier part of the session will be retained in the separate files with the original name(s).

Alternatively, at the end of a session you can rename SPSS.LIS on the hard disk (and you will probably then wish to copy the renamed file to a floppy disk.

To rename SPSS.LIS use a **DOS** 'ren' (= 'rename') command, such as:

```
C>ren SPSS.LIS  output1
```

This would rename the existing listing file, giving it the name 'output1' (later listing files might be named output2, output3, etc.). You can use any name you like (8 letters or less as a forename, with an optional three-letter surname).

You might then copy the renamed listing file to a floppy disk in, say, *Drive A*, with the command:

```
C>copy output1 a:
```

This would copy 'output1' to a file of the same name on *Drive A*.

The file can be edited or printed. To print the file directly from **DOS** you could simply issue the command:

```
C>type spss.lis>prn
```

In practice, however, you would probably want to edit the 'raw' listing file (perhaps with a word-processor) before printing annotated extracts.

Finally, it is worth noting that "SPSS.LIS" is actually produced throughout the session, not just at the end, and may be edited (with **REVIEW**) at any time during the session by issuing at the SPSS/PC: prompt either:

```
SPSS/PC: review "spss.lis".
```

or

```
SPSS/PC: review listing.
```

This will produce a display of the bottom part of the listing file. You can then use the various cursor movement controls (including **<Pg Up>** and **<Pg Dn>**) to move to that part of the file which you wish to read or edit.

7 *Statistical Analysis II: MEANS and T-TEST*

In this chapter we continue our examination of SPSS/PC+ procedures by introducing two more procedure commands – **MEANS** and **T-TEST**.

7.1 The MEANS procedure

The **MEANS** procedure groups cases together on the basis of their values on one or more independent variables and then displays group statistics for any specified dependent variables. Thus cases might be grouped in terms of their values on the variable *sex*. The mean values (averages) and standard deviations would then be computed for specified variables for each of the groups formed in this way (in this case, for males and females).

In the Gender Attitudes Study, the variable *prs04* is the respondent's answer ('true' or 'false') to the statement **'I have to admit that I get angry quite often'**. It is thus a categorical variable which has just two valid values (coded 1 for 'true' and 2 for 'false'). The command:

```
SPSS/PC: MEANS var001 to var005 BY prs04.
```

would produce five tables, one for each of the variables specified to the left of the **BY** keyword. In each table the means and standard deviations for one of the dependent variables (*var001*, *var002*, etc.) would be given for each of the two groups (one group for each of the two values of *prs04*). A count of the number of cases in each group would also be provided, and the same statistics would also be given for the population as a whole. The dependent variables (those to the left of the **BY** keyword) must be numerical (rather than categorical) if the averages and standard deviations are to be at all meaningful. The independent variables (those to the right of the **BY** keyword), on the other hand, will usually be categorical in nature (like *sex* or *prs04*). From this command:

```
SPSS/PC: MEANS var001 to var010 BY sex.
```

we would obtain ten tables of the means, standard deviations and valid counts for variables *var001* to *var010* for males and females separately. The above command produced the following output for *var008*:

```
Summaries of    VAR008      Men more sexy
By levels of    SEX         Subject sex

Variable          Value  Label              Mean     Std Dev      Cases

For Entire Population                     4.7600      1.8905         50

SEX                   1  Male             3.8636      1.8334         22
SEX                   2  Female           5.4643      1.6439         28

 Total Cases  =  50
```

It is possible to break down the population by more than one variable (i.e. to use more than one *independent* variable). Thus the command:

```
SPSS/PC: MEANS var001 to var008 BY sex BY prs01.
```

would produce eight separate tables, one for each of the dependent variables *var001* to *var008*. Each table would include statistics for the population as a whole, for the male group and the female group separately, and for each of the two subgroups (*prs01* value 1, and *prs01* value 2) *within each sex group* (i.e. four subgroups altogether). Thus each possible combination of values of the independent variables defines a 'group' or 'subgroup'. The variable *prs01* is coded on the basis of whether the respondent agrees ('true', coded 1) or disagrees ('false', coded 2) with the statement **'I often feel a little depressed'**. The above command produced the following as one of the eight output tables:

```
Summaries of    VAR007      Women more sensitive
By levels of    SEX         Subject sex
                PRS01       Depressed often (SELF)

Variable          Value  Label       Mean     Std Dev     Cases

For Entire Population              4.0400      1.8290        50

SEX                   1  Male      4.0000      1.9518        22
  PRS01               1  TRUE      4.4000      1.9551        10
  PRS01               2  FALSE     3.6667      1.9695        12

SEX                   2  Female    4.0714      1.7623        28
  PRS01               1  TRUE      3.6842      1.9164        19
  PRS01               2  FALSE     4.8889      1.0541         9

  Total Cases =          50
```

An examination of this table suggests (although further statistics would be needed to support this interpretation) that while there is no overall difference in the extent to which males and females believe that women are more sensitive than men, males who get depressed are less likely to believe this ('women are more sensitive') than males who do not get depressed. On the other hand, females who get depressed seem somewhat *more* likely to believe that females are sensitive than are females who do *not* get depressed.

Although the three variables we have used as examples of independent variables (*sex*, *prs01* and *prs4*) each split the sample into just two groups, other independent variables might split the sample into, for example, three age-groups, four faculty groups, five social class groups, etc.). Taking into account the fact that up to five **BY** words can be used in any command, it becomes clear that the number of subgroups that can be formed using a **MEANS** command may be very large indeed. As more subgroups are specified, however, the average number of cases which fall into each subgroup will diminish, and the numbers within particular groups might soon become so small that the standard deviations, in particular, will become meaningless.

The order in which independent variables are named affects the nature of the display of the table (but not the values contained within it). The first-named independent variable becomes the 'major' classification criterion, with later-named independent variables becoming increasingly 'minor'.

Two or more lists of variables can be included on separate subcommands within a single **MEANS** command. For example:

```
SPSS/PC: MEANS var001 to var004 BY prs04 BY sex
       : /var007 var009 var010 by st06.
```

This example would produce seven separate tables (four requested by the first subcommand and three by the second).

The command:

```
SPSS/PC: means var001 to var003 BY prs01 prs02 prs03.
```

will produce nine separate tables – var001 with prs01
var001 with prs02
var001 with prs03
var002 with prs01
etc.

The default display presents means, standard deviations and counts for the entire population, and for each group and subgroup. The output is fully labelled with variable labels and value labels. If a case has a missing value for one or more of the variables involved in a requested table then the case is omitted from all of the statistics calculated for the table.

An **Options** subcommand and/or a **Statistics** subcommand can be used to tailor the information provided in the display table.

The **Options** available for the **MEANS** procedure relate to the use of missing data, the suppression of names and labels, and the display of group sums and group variances. **Options** are specified by a number. For example, **Option 1** requests that missing cases be included in the calculations, and **Option 3** causes all labels to be suppressed.

A **Statistics** subcommand can also be used. Just two **Statistics** specifications are available, and they are also requested by number. **Statistic 1** requests that a one-way analysis of variance (see Chapter 12) be performed using the first named independent variable.

Statistic 2 requests a test of linearity. This again involves a one-way analysis of variance using the first-named independent variable, and linearity is assessed.

A full **MEANS** command might take the form:

```
SPSS/PC: means var002 to var006 BY sex BY prs06
       : /var002 var004 var007 BY st03
       : /options= 3
       : /statistics= 1.
```

This requests that eight tables (5 + 3) be displayed without labels (**Option 3**) and that a one-way analysis of variance be performed on the data within each table (**Statistic 1**).

7.2. The T-TEST procedure

The **T-TEST** procedure is used to compare the means (average values) of two variables for a single group of cases, or of the same variable for two different groups. It should be used with numeric data and *not* with coded variables. Unlike the **MEANS** procedure, which produces only a table of descriptive statistics (unless either of the **Statistics** options is specified), the **T-TEST** procedure performs a statistical test. The *t*-value obtained is an inferential statistic which allows us to draw conclusions (make inferences) about groups of cases or about variables represented in the dataset.

There are two forms of the *t*-test.

(a) **An independent samples *t*-test** is used to compare two groups on a single specified variable. Such a test can reveal, for example, whether the *age* scores of our male subjects and our female subjects are significantly different.

Although a coded variable, *sex*, is involved in this comparison, the coded values are used here only to define the two groups. Apart from using the codes to form the groups, the system makes no use of the values 1 for 'male' and 2 for 'female' in the statistical analysis. The computation of the *t*-test statistic involves only the numbers that represent *age* in years.

(b) A **paired samples *t*-test** (sometimes referred to as a 'paired *t*-test', or a 'correlated ***t*-test') is used to assess whether, for a single group of subjects, the score on one variable is significantly different from the score on another variable. Thus it could be used to determine whether respondents' values for** *var001* (the 1–7 agreement rating supplied in response to the statement **'Women are "equal" to men in every way'**) are significantly different from the same respondents' values for the variable *var010* – reflecting their level of agreement with the statement **'Most men feel very threatened by highly intelligent women'**. The data relating to *var001* and *var010* for the same respondent are specifically paired in the calculation that produces a *t*-value. It should be noted that some statisticians object to the use of 7-point scale data in such analyses although others are satisfied that such data need not infringe the assumptions of the *t*-test.

When a *t*-value has been calculated we need to know the associated probability. Low *t*-values may reflect the effects of mere chance. High *t*-values are more likely to reflect a 'real' difference between the groups (or between the variables).

The probability value associated with any particular *t*-value will depend on how many cases the calculation involved (this is directly related to the 'degrees of freedom' value which is actually used in calculating the *t*-value) and on the nature of the hypothesis being tested. If the hypothesis is *directional* (e.g. males are older than females *or* females are older than males) then a lower value will be needed to obtain a significant probability value than if the hypothesis is of a *non-directional* kind (e.g. there is a difference, in one direction or the other, between the ages of males and females). A directional hypothesis is tested with a *one-tail test* and a non-directional hypothesis with a *two-tail test*. SPSS/PC+ output includes two-tail probabilities. The equivalent one-tail probability is found simply by dividing the two-tail probability by 2.

The independent t-test

The SPSS/PC+ procedure command for an **independent *t*-test** starts by specifying how two comparison groups are to be formed. Thus if we wish to form two separate groups, one consisting of the male respondents and the other consisting of the female respondents, we must specify that the grouping variable is *sex* and then instruct the system how to form groups using that variable.

Thus to request an independent *t*-test to compare *age* values for the two sexes, we would use the command:

```
SPSS/PC: t-test groups=sex (1,2) / variables = age.
```

The first part of this command (before the slash, /) instructs the program to define two groups on the basis of the variable *sex*. Cases with the value 1 on the variable *sex* (in this case males) will form one group and those with the value 2 (in this case females) will form the other group. If only one value is specified within the brackets then cases with this value form one group and all other cases are treated together as a second group.

The second part of the command (after the slash, /) tells the program to compare the two defined groups on the variable(s) listed (in this case the single variable *age*).

Here is another example of an independent *t*-test command:

```
SPSS/PC: t-test groups= prs06 (1,2)
       : / variables = age var006 to var008.
```

This would lead to the formation of two groups, one consisting of those who said (*prs06* value 1) that they enjoy taking risks, and one consisting of those who said that they didn't (*prs06* value 2). These two groups would then be compared on the variable *age*, and on their responses to three of the attitude questions (*var006* to *var008*).

The above command revealed that the group of respondents who said that they enjoyed taking risks differed in their degree of agreement with the statement **'Women are emotionally more sensitive than men'** (*var007*) from those who said that they did *not* enjoy taking risks (*t*-value = 2.62, two-tailed probability = .012).

The output which revealed this finding took the following form:

```
Independent samples of  PRS06     Enjoy risks (SELF)

Group 1:  PRS06  EQ 1           Group 2:  PRS06  EQ 2

t-test for:  VAR007    Women more sensitive

                Number                Standard    Standard
               of Cases     Mean     Deviation      Error

     Group 1      28       4.6071       1.707        .323
     Group 2      22       3.3182       1.756        .374

               : Pooled Variance Estimate :     Separate
               :                          :     Variance
   F    2-Tail :    t   Degrees of 2-Tail :     Estimate
 Value  Prob.  :  Value   Freedom   Prob. :
               :                          :   (see below)
  1.06   .878  :   2.62      48      .012 :
```

The default output from the independent *t*-test procedure (as above) displays the means, standard deviations, standard errors and counts (number of cases) of valid cases for each group. In addition, the degrees of freedom are given and there are two *t*-values, one based on the 'pooled variance estimate' and one based on the 'separate variance estimate'. This is explained below. Each of the *t*-values is then expressed in terms of its probability – the two-tailed probability of the difference between the two group means. The one-tailed probability is simply half of this value – for the above example it would be .006. Cases which have missing values for the dependent variable are *not* included in the calculation (in the example above there were no missing cases). The display also includes the variable labels.

A 'variance ratio' (*F*-test) value is given, along with its probability value. The *F*-test compares the variances of the two samples to determine whether these variances are significantly different. The *F*-test result is used in the following way. If the *F*-value is significant (i.e. associated with a probability of *less than* .05) then the *t*-value calculated from the **separate variance estimate** should normally be used (the relevant information, which *would* be included in normal SPSS/PC+ output is *not* included in the above table). If, however, the *F*-value is *not* significant then the *t*-value calculated from the **pooled variance estimate** should normally be used. (In most cases the two *t*-values are in fact very similar.) Thus in the above example the *F*-value (1.06) is *not* significant (probability = 0.878) and we can therefore conclude that the two variances are *not* significantly different. Thus the calculation based on the pooled variance estimate is used to draw a conclusion about whether the means of the two groups differ significantly. In this case we take 2.62 to be the relevant value of *t* and we base our conclusions on that.

Those who are relatively new to statistics may find some of this discussion difficult. A fuller (and gentler) explanation of the independent *t*-test will be found in most introductory statistics textbooks.

Only one **Groups** subcommand and one **Variables** subcommand can be included in a **T-TEST** command (although the **Variables** subcommand may specify up to 50 variables for analysis). No more than 50 independent *t*-tests can be requested by a single **T-TEST** command.

A number of **Options** are available for the independent *t*-test.

Option 1 specifies that missing values should be *included* in the analysis (they are normally excluded, of course).

Option 2 specifies that a case (subject) should *not* be included in *any* of the *t*-test analyses specified by the command if that case has missing data for *any* of the variables specified on the *t*-test command. Thus with the following instruction:

```
SPSS/PC: t-test groups= sex (1,2)
       : / variables= var001 to var006
       : / options= 2.
```

any case with a missing value for any one or more of the listed variables would be excluded from *all* of the six analyses requested by the command.

The effect of **Option 2** can be illustrated by running the above command, using the Gender Attitudes Survey data, once *with* the **Option 2** subcommand and once *without* it. In either case, each output display will include a report of how many male and female cases have been processed for the particular analysis. With **Option 2** requested the number of cases analyzed remains constant throughout the series of six displays, each analysis being based on 21 male and 26 female cases. These are the cases within the full dataset (of 22 males and 28 females) for which complete valid data are available for all of the variables *var001* to *var006*. When no **Options** subcommand is issued, however, the numbers of cases analyzed over the six *t*-tests requested vary between 21 and 22 for the males and 27 and 28 for the females. Each analysis is based on all cases of the which have a valid value for the particular variable being analyzed.

This latter strategy of deleting only those cases for which data are missing for a variable directly involved in the specific analysis being performed is known as **pairwise deletion**. This is usually the SPSS/PC+ default strategy for dealing with analyses in which one or more cases have missing values for a specified variable. The alternative strategy of deleting a case from *all* analyses in a listed specification if it is missing valid data for *any* of the variables in the list, is known as 'listwise deletion' and can be requested by specifying **Option 2** within a **T-TEST** command.

Option 3 specifies that variable labels should *not* be included on the output.

No additional **Statistics** are available for the **T-TEST** procedure.

The following limitations apply to the **T-TEST** procedure when independent *t*-tests are requested and should be borne in mind if multiple analyses are needed:

- Only *one* **Groups** subcommand can appear within a single **T-TEST** command, and the **Groups** subcommand can name only one variable (however, a sequence of several **T-TEST** commands with different **Groups** subcommands can of course be submitted to the system).
- Only one each of the **Variables** and **Options** subcommands can be included within a single independent **T-TEST** command (although the analysis of several variables may be requested within a single **Variables** subcommand, and a number of options can be specified within a single **Options** subcommand).

- No more than 50 variables can be specified on single **T-TEST** command (and they should all be numerical rather than categorical).

The paired t-test

The **paired *t*-test** is used to determine whether, for a single group of subjects, the score on one numerical variable is significantly different from the score on another numerical variable.

The command for a **paired** *t*-test takes the form:

```
T-TEST Pairs = var001 WITH var002.
```

Thus a paired *t*-test could be used to assess whether the level of agreement expressed by all of the respondents towards one statement (as represented, for example, by *var001*) is significantly different from that which all of these respondents expressed towards another statement (for example, that represented by *var010*).

The **T-TEST** command used to specify such a comparison would be:

```
SPSS/PC: t-test pairs= var001 with var010.
```

The keyword **WITH** can be omitted if you wish to perform paired *t*-tests between every pair in a variable list. Thus:

```
SPSS/PC: t-test pairs= var001 var010.
```

is equivalent to the command immediately above. However, the command:

```
SPSS/PC: t-test pairs= var001 to var006.
```

will produce 15 paired *t*-test comparisons, because each variable, *var001* to *var006*, will be paired with every other variable on the (implied) list.

Lists of variables can be included in the command on one or both sides of **WITH**:

```
SPSS/PC: t-test pairs= var001 to var004 with var006 to var010.
```

Here each of the four variables in the first part of the list will be paired with the variables *var006* to *var010* (making 20 paired *t*-tests in all). Note that in using this command the differences between pairs of the first four variables will *not* be calculated.

Multiple **Pairs** statements can be included within a single **T-TEST** command. Separate statements are separated with a slash (/) and the **Pairs** keyword need not be included after the first time. Thus:

```
SPSS/PC: t-test pairs= var001 to var006 with var008
       : /var001 with var002
       : /var003 with var005 var007 to var010.
```

This would yield 12 paired *t*-test analyses altogether.

The above command produced, among other findings, confirmation of the directional (one-tail) hypothesis that respondents would agree significantly more with the statement **'More legislation is needed to ensure equal rights'** (*var002*) than with the statement **'Generally speaking, men have a higher level of sexual desire than women'** (*var008*). The average values from the two 7-point scales was 2.10 for *var002* and 4.76 for *var008*. Because of the scoring method used, a low value indicates more agreement than a high value. The data from the 50 respondents yielded a *t*-value of 6.62 which corresponds to a one-tailed probability of less than .001. The difference is therefore in the direction predicted, and is highly significant.

The output which revealed this finding took the following form:

```
Paired samples t-test:    VAR002        New  legislation  needed

                          VAR008    Men more sexy

Variable    Number                  Standard   Standard
           of Cases      Mean      Deviation     Error

VAR002        50        2.1000        1.644       .233
VAR008        50        4.7600        1.890       .267

(Difference) Standard  Standard       2-Tail    t   Degrees of  2-Tail
    Mean    Deviation   Error    Corr. Prob.  Value   Freedom    Prob.

   -2.6600      2.840    .402   -.288  .043  -6.62      49        .000
```

By default, any cases with missing values for either variable are not included in the calculation of a paired *t*-test. The default output from the procedure (as above) includes the variable labels for the two variables under examination and displays the mean difference of the pairs together with the standard deviation and standard error of this value. A correlation coefficient is calculated between the two variables and is printed together with its two-tail probability value. The output then includes the *t*-value, and the degrees of freedom and two-tailed probability associated with this value. This last statistic is an estimate of the probability that the difference found between the means of the two variables might be due to chance alone. As we can see this is given as '.000'. We would report that the difference between the means was associated with a (chance) probability of 'less than .001'. Another way of expressing this would be to say that our finding would occur by chance less than one in 1000 times. Or, we might say that the finding is 'significant at the .001 level'.

The **Options** available for the independent *t*-test are also available for the paired *t*-test.

Option 1 specifies that missing values should be *included* in the analysis.

Option 2 specifies 'listwise deletion' of cases with missing data (this is explained in the section relating to the independent *t*-test, p. 98).

Option 3 specifies that variable labels should *not* be included on the output.

There is no **Option 4** for the **T-TEST** command.

Option 5, which can be used within a **T-TEST** command only when a paired *t*-test is requested, specifies that a special pairing interpretation should be applied the variables listed. If this option is used, the command:

```
SPSS/PC: t-test pairs= var001 to var005 with var006 to var010
       : /options = 5.
```

leads not to 25 comparisons (5 X 5) but to *five* only. The first variable before the keyword **WITH** is compared to the first variable after **WITH**, the second before is compared with the second after, etc., until finally the last (fifth) variable before **WITH** (*var005*) is compared with the final (fifth) variable after **WITH** (*var010*).

No additional **Statistics** are available for the **T-TEST** procedure, and there is no SPSS/PC+ limit on the number of variables which can be compared with a paired *t*-test procedure command (however, there can be no more than 500 variables in an active file).

8 *Transforming Variables; Selecting and Sorting Cases*

Once a file containing the raw data has been established, the user may wish to transform the data in certain ways. Thus the coding system originally used for a variable may be inappropriate for certain types of analysis. For example, if a variable *age* has been coded in terms of years it will not be possible to directly compare a young group with a middle-aged or old group. With a **RECODE** command, however, the data in the active file can be transformed so that *age* (in years) is recoded into, say, 'young' – up to 25 years; 'middle' – 26 to 50 years; and 'old' – over 50 years. Such a re-coding would produce three age groups rather the original groupings (one for each separate age).

Sometimes the user may wish to create new composite variables (say, the sum of *var001*, *var002* and *var003*), and this can be achieved by using another data transformation command, **COMPUTE**.

In this chapter we will first consider the use of the data transformation commands **RECODE** and **COMPUTE**, before discussing **conditional transformations** (involving the command **IF**). We will then consider how particular cases can be selected for analysis, and how the order of cases within a file may be changed using the **SORT** command.

8.1 The RECODE command

A **RECODE** command specifies variable(s) to be recoded and provides an equation (=) for transforming the *old* values into *new* values (in this order):

The following examples illustrate how data may be recoded:

```
SPSS/PC: RECODE var001 (1,2,3=1) (4=2) (5,6,7=3).
```

This transforms values from the original 7-point scale into just three values: 'low' (the old codings below 4 are now coded together as 1), 'middle' (the old 4 becomes 2) and 'high' (the old codings above 4 are now coded together as 3). The system takes note of the **RECODE** instruction but the actual process of recoding is not implemented until a subsequent procedure is requested.

If you issue the above command on the data from the Gender Attitudes Survey, follow it with a request for a **FREQUENCIES** display:

```
SPSS/PC: frequencies var001.
```

The system will recode the variables and there will be a delay before the **FREQUENCIES** display is provided. This display will then indicate that values have been recoded so that all valid data is now within the range 1 to 3. You will notice that the missing data (coded 9) remain unchanged. Because there are now no valid data with the value 7, the label associated with that value ('Strongly Disagree') is no longer included in the display. This points to the need, after a recode, to revise the existing value labels (a new **VALUE LABELS** command can be issued in response to the SPSS/PC: prompt and a new set of labels supplied – these will override any existing value labels for that variable).

The keywords **THRU**, **LOWEST** (also written as **LO**), **HIGHEST** (also written as **HI**), and **ELSE** can be used. Lists of variables can be included within the **RECODE** command and such lists can include the keyword **TO**:

```
SPSS/PC: recode salary (LO thru 5000=1)
       : (5001 thru 7500=2) (7501 thru hi=3).
```

This recodes the initial *salary* values to form just three *salary* groups. Valid values of *salary* will now be within the range 1 to 3.

The keywords **LO**, **HI** and **ELSE** recode a missing value if it is within the range specified. Thus in this example any missing data for *salary* (originally coded 99999) will now have the value 3 (and hence be indistinguishable from high valid salary values).

The next example:

```
SPSS/PC: recode var002 to var005 (LO thru 4=1) (ELSE = 2).
```

recodes the initial values for the four named variables to form two groups for each. In this example any missing data for the variables (originally coded 9) will now have the value 2 (and hence be indistinguishable from high valid values).

However, in the following example:

```
SPSS/PC: recode var004 var006 to var010 (1,2,3=1)
       : (4=2) (5,6,7=3)
```

the missing value (9) would be unaffected.

Having changed coding values you will need to change the relevant value labels (and possibly the variable label).

If the original values for a variable have been collapsed (say the values for *age* – in years – have been collapsed into just three categories) then the fuller raw data will not be accessible during the remainder of this session. There is no way of recoding the three *age* categories to regain information about the respondents' actual ages (remember 'you can't unscramble eggs'). The original file will have to be included again (using the **INCLUDE** command) for the original values to be made available.

8.2 The COMPUTE command

The **COMPUTE** command creates a new variable using the data from the original dataset. Thus:

```
SPSS/PC: COMPUTE sum  = var001 + var002.
```

This would, for each case, compute a value for the new variable *sum* by adding the values for two existing variables – *var001* and *var002*. Thus *new variables can be created without adding new data.*

The arithmetic operations allowed in compute commands are: addition (+); subtraction (–) ; multiplication (use *) ; division (use /) and exponentiation (i.e. squares, cubes, etc.) (use **).

The example:

```
SPSS/PC: COMPUTE monsal = salary/12.
```

would compute the expected monthly salary (*monsal*) by dividing the (annual) *salary* by 12.

This would not be the case, however, if *salary* had been recoded, as in the example given in the previous section, to a range of 1 to 3. If *salary* had been recoded in this way then the above command would divide the *recoded* values by 12, and *monsal* would clearly *not* be a direct representation of the monthly salary. To find the monthly salary we would have to retrieve the original codings in the dataset by including the original file again. Such an operation, of course, will restore *all* of the original values and thus cancel the effects of any **RECODE** and **COMPUTE** commands that might have been issued since the file was last made active.

The following **COMPUTE** command:

```
SPSS/PC: COMPUTE wt= (wst * 14) + wlb.
```

would create a new variable *wt* (total weight in pounds) from two original variables – the weight in stone (*wst*) and the residual weight in pounds (*wlb*).

The command:

```
SPSS/PC: COMPUTE ht= (htf * 12) + hti.
```

would create a new variable *ht* (total height in inches) from two original variables – the height in feet (*htf*) and the residual height in inches (*hti*). This example and the previous one show how brackets can be used to specify the order of operation. Remember:

`A * (B + C)` is **not** equivalent to `(A * B) + C`

e.g. if A = 2, B = 3 and C = 4 then A*(B+C) = 14 and (A*B)+C = 10.

Once a computed variable has been added to the active file, it can be treated like any other variable. Thus it can be provided with a variable label and value labels, it can be recoded, and it can be used to compute another new variable.

Thus, using the variables *wst* and *htf* created above by the use of **COMPUTE** commands, we could derive an index of weight relative to height (let's call the variable *relsize*) by dividing the total weight in pounds by the total hcight in inches. Thus:

```
COMPUTE relsize = wt / ht.
```

We can also use **COMPUTE** to duplicate an existing variable:

```
COMPUTE varnew = varold.
```

The copy of the original variable could now be subjected to a **RECODE** command, and thus we would have two copies of the original variable, one retaining the original values and one with recoded values.

Various transformation functions (including log functions, rounding, sines, etc.) can be used within **COMPUTE** commands (see the *SPSS/PC+ 4.0 Base Manual*). We might, for example, wish to round values for the computed variable *relsize* to that all values are whole numbers ('integers'). If we had already created *relsize*, the appropriate command to bring about such a rounding transformation would be:

```
COMPUTE relsize = rnd (relsize).
```

although we could create a 'rounded' variable in a single step, using:

```
COMPUTE relsize = rnd (wt / ht).
```

COMPUTE can also be used to change the value for a variable in a uniform way, for example to double the value or to add a constant sum to the existing value. This can be done without the need to create a new variable. In effect, the existing variable is re-created with the revised value. This is done by specifying the variable on both sides of the equation, as in:

```
SPSS/PC: COMPUTE oldvar = oldvar * 2.
```

which doubles the initial value of the variable for every case in the active file, and:

```
SPSS/PC: COMPUTE oldvar = oldvar + 10.
```

If a subject has a missing value for one or more of the variables used in creating a new variable, then the new variable is also treated as missing for that subject. Since it is declared to be missing by the system (rather than by the user) it is said to be **system-missing**.

COMPUTE is a transformation command, and only one new variable can be created with a single **COMPUTE** command. Information is noted by the system as the command is issued, but values for the new variable are not created until the next procedure command is issued. The newly created variable then become part of the active file and contributes towards the limit of 500 variables within that file.

Using COMPUTE to score tests

One special use of **COMPUTE** is to derive a composite score from a number of related or similar variables.

Suppose, as in the Gender Attitude Survey, we have a twelve-item personality test, and wish to derive scores on two dimensions (in this case 'extraversion' and 'neuroticism'). The test given to respondents in this survey was devised by Hans Eysenck (1958). From the scoring key supplied with this test, we know that six of the items from the personality questionnaire are used to derive an extraversion score, while the other six items are used to derive a neuroticism score. The data relating to the items have been coded as the 12 variables *E1* to *E12*. In each case the code 1 has been entered for a 'yes' response, 2 for a 'not sure' response', and 3 for a 'no' response.

From the instructions on how the test should be scored, we learn that a 'no' response to any of the six extraversion items *lowers* the total extraversion score by 1 point, whereas a 'yes' response *increases* the score by 1 point. 'Not sure' responses have no effect on the score. Exactly the same procedure is followed in deriving a total neuroticism score from the responses to the six neuroticism items.

Thus one way to compute total extraversion and neuroticism scores for each of the respondents in the Gender Attitudes Survey would be first to recode the original data and then to compute a total extraversion score and a total neuroticism score for each respondent by adding their recoded data values for the relevant items.

In devising a suitable **RECODE** command, we might decide that the 'yes' codes (1) do *not* need to be changed. The 'not sure' responses (originally coded as 2) could be re-coded to 0, and the 'no' responses (originally coded as 2) could be recoded as -1. These changes could be made using the single **RECODE** command:

```
SPSS/PC: RECODE E1 to E12 (2= 0) (3 = -1).
```

A subsequent **LIST** command, issued to examine the effect of this transformation:

```
SPSS/PC: list id e1 to e12.
```

would produce the following display:

```
ID E1 E2 E3 E4 E5 E6 E7 E8 E9 E10 E11 E12

 1  1  1 -1 -1  1 -1  1  1  0   1  -1  -1
 2  1  1 -1  0 -1  1  1  1 -1   0   1   1
 3  1  1  1  1  1  1  0 -1  1   0   1  -1
 4  1 -1  1  0 -1  1 -1 -1  0   1   1   1
 5  1 -1  0 -1  1  1  1 -1  1   1   0   1
....
47  1  1  1  1 -1  1  1  1  0   1   1   1
48  1  1  1  1  1  1  1  1 -1   0   1   1
49  1  1  1  1 -1  1  1  1 -1   1   1   1
50  1 -1 -1  0 -1  1  1 -1  1   0   1  -1

Number of cases read =    50    Number of cases listed =    50
```

The extraversion score (*extra*) could then be computed from the recoded values of the six relevant variables:

```
SPSS/PC: compute extra = (E2 + E4 + E7 + E8 + E10 + E12).
```

and a neuroticism score (*neur*) could be computed from the other six E variables:

```
SPSS/PC: compute neur = (E1 + E3 + E5 + E6 + E9 + E11).
```

A suitable **LIST** command would now display the values of the new variables:

```
SPSS/PC: list id extra neur.
```

```
 ID     EXTRA     NEUR

  1      2.00    -1.00
  2      4.00     0.0
  3      0.0      6.00
  4     -1.00     3.00
  5      0.0      4.00
 ....
 47      6.00     3.00
 48      5.00     4.00
 49      6.00     2.00
 50     -2.00     2.00
```

A **FREQUENCIES** display for the new variables *extra* and *neur* shows that the mean for *extra* is 1.860 and the standard deviation is 2.799. The mean for *neur* is 2.140 and the standard deviation is 3.064.

Using the above strategy, and issuing the **RECODE** and **COMPUTE** commands interactively, the new variables *extra* and *neur* would be created for the current session only. The computed variables could be specified in any further procedure commands as long as the session lasted, but at the end of the session these variables would no longer exist.

A further example of test scoring using **COMPUTE** involves the variables *var001* to *var010*, relating to 10 attitude statements from the Gender Attitude Questionnaire. Respondents were asked to indicate how far they agreed or disagreed with each statement by circling one number on a 7-point scale from 1 to 7, where 1 was labelled 'Strongly Agree' and 7 was labelled 'Strongly disagree'.

An examination of these statements might lead us to suggest that agreement with some of them would reflect a 'non-sexist' viewpoint, while agreement with others would reflect the opposite, a 'sexist' outlook. Thus the data relating to an individual's responses to the set of statements could be used to assess his or her degree of 'sexism'. (There would probably not be total consensus among judges about which statements are 'sexist' and which 'non-sexist', and in practice the construction of a test would not depend on such subjective judgements. See Chapter 14.)

If we judge statements 1, 2 and 9 to be 'non-sexist' this would imply that for these items a *high* score (indicating *disagreement*) would indicate a *high* degree of *sexism*. On the other hand, statements 3, 4, 5 and 8 might be judged to be 'sexist', and this would imply that *agreement* with these statements (indicated by a *low* coded value) would reflect a sexist attitude. Thus for these items a low value should *increase* the *sexism* score. The remaining three statements might be regarded as somewhat ambiguous with regard to their 'sexist' or 'non-sexist' nature, and they might therefore be omitted from the process of deriving a composite *sexism* score.

To obtain the overall *sexism* score for each respondent, we would need to combine information from both the 'positive sexism' items and the 'negative sexism' items. We could do this by adding together the scores obtained for each of the 'positive sexism' items and adding the *reversed* scores for the 'negative sexism' items. Reversing the score means that a 7 would become 1, a 6 would become 2, etc. The general formula for reversing scores on a 7-point scale is '(8 – original score) = reversed score'. Thus for an original score of 3, the reversed score is 8–3 = 5.

The appropriate formula for deriving a single *sexism* score for each respondent from their responses to the seven relevant statements (including a reversal of the original code for each of the four 'sexist' statements) could be specified within a single **COMPUTE** command. Thus:

```
SPSS/PC: compute sexism = (var001 + var002 + var009 +
       : (8 - var003) + (8 - var004) + (8 - var005) +
       : (8 - var008)).
```

Here a high *sexism* score would be obtained by having a *low* score for variables *var003*, *var004*, *var005* and *var008*, and by having a *high* score for variables *var001*, *var002* and *var009*. Agreement with any of the 'sexist' items (originally coded as a *low* value) has now been made to increase the *sexism* score substantially, through the application of the '8 – original score' formula to the original coding.

The new *sexism* variable would have a 'theoretical range' from 7 to 49. The actual range for the respondents in the Gender Attitudes Survey (as shown by a **FREQUENCIES** display) is 14 to 38. The mean *sexism* score for this population is 24.229 and the standard deviation is 5.846.

Lowering the derived *sexism* scores by 7 points (i.e. one point for each item) would allow those respondents who were totally 'non-sexist' (i.e. those who strongly agreed with each of the 'non-sexist' items and strongly *dis*agreed with each of the 'sexist' items) to obtain a zero score. Such a lowering of the *sexism* scores by 7 points across all respondents could be achieved by including an additional **COMPUTE** command:

```
SPSS/PC: compute sexism = sexism - 7.
```

The theoretical range of this *sexism* score would now be 0–42. The actual range for the Survey respondents was 7 to 31. This was revealed by a new **FREQUENCIES** display but could also have been calculated from the previous range (i.e. by subtracting 7 from both the previous minimum and the previous maximum). The mean is lowered by exactly 7.00 (it is now 17.229). The standard deviation is never affected by a uniform addition or subtraction of any value to all of a set of scores, and therefore remains at 5.846.

Again using the Gender Attitudes Survey data, we could compute a *problem* score by combining information from several of the *prs* variables (each of which is coded 1 for 'true' and 2 for 'false'). This could be achieved by adding one point for each of a range of items which the respondent reported as a frequent occurrence ('often bored', 'often depressed', 'often anxious', etc.). Allowing for the fact that denial of such problems has been coded as 2 ('false'), and admission of such problems as 1 ('true'), the overall *problem* score could be computed in the following way:

```
SPSS/PC: compute problem = (2 - prs01) + (2 - prs03) +
       : (2 - prs04) + (2 - prs07) + (2 - prs08) +
       : (2 - prs09) + (2 - prs10).
```

This would produce a score, for each individual case, ranging from 0 ('no problems') to 7 ('many problems'). For the Gender Attitudes Survey data the mean value for *problem* is 2.857 and the standard deviation is 1.893.

8.3 Conditional transformations: IF

The **COMPUTE** and **RECODE** transformations described so far affect all cases in the active file. Sometimes, however, it is useful to transform a variable only for those cases in which certain conditions are fulfilled. Such transformations are known as 'conditional transformations' and they affect any particular case only **IF** a specified logical expression is *true* for that case.

The examples given in this section describe a hypothetical dataset. They do not refer to the Gender Attitudes Survey data.

We might wish to assign a value 1 for a new variable *pass* only for those cases in which an examination mark (the value of a variable *test*) reaches a criterion 'pass mark'. A suitable conditional transformation would assign a specified value for the new variable only to those cases for which the value of *test* is *greater than* a specified criterion value. Such a command could take the form:

(a) `SPSS/PC: IF (test GT 49) pass = 1.`

GT is a **relational operator** which stands for *greater than*. The above command would assign the value 1 for the new variable *pass* to any case for which the logical expression '(test GT 49)' is true, i.e. for any case for which the *test* value (e.g. an examination mark) is more than 49. For those cases in which the logical expression was evaluated as 'not true' (or for which it could not be evaluated) the system would assign a **system-missing** value to *pass*.

We might wish to assign another value to *pass* for those students who had actually failed the examination. For example, we could specify that all students who obtained marks less than or equal to 49 (**LE 49**) be assigned the value 0 for the variable *pass*. This could be done in either of the following ways (the relational operator **LE** stands for 'less than or equal to'):

(b) `IF (test LE 49) pass = 0`

or

(c) `IF (NOT test GT 49) pass = 0.`

We see here that the keyword **NOT** (a **logical operator**) can be used to reverse the 'true' or 'false' status of an expression.

The effect of issuing the two commands ((a) together with either (b) or (c)) will be to assign either the value 0 or the value 1 to all cases with a valid *test* score. Cases with a *test* value below 50 will be assigned a 0 for *pass*. Those with 50 or more marks will be assigned the value 1 for *pass*. The logical expression will not be able to be evaluated for any case with a missing value for *test*, and any such case will be assigned a system-missing value for *pass*.

Many different relational operators are permitted in logical expressions, and their various effects are outlined in the table opposite.

Relational operators
The expression (X **EQ** Y) is true if X is *equal to* Y
The expression (X **NE** Y) is true if X is *not equal to* Y
The expression (X **GT** Y) is true if X is *greater than* Y
The expression (X **GE** Y) is true if X is *greater than or equal to* Y
The expression (X **LT** Y) is true if X is *less than* Y
The expression (X **LE** Y) is true if X is *less than or equal to* Y

Relational operators can be written within commands either in character form as above (GE, LT, etc.) or in symbol form (thus EQ could be replaced by = , GT could be replaced by > , etc.).

The **X** and **Y** components in the table need *not* stand for simple variables or numeric values. Either or both of the components can be complex expressions themselves. Thus:

```
SPSS/PC+ IF (((v1/v2) * v3) NE ((v4 - v5) * v8)) index = 0.
```

Here the **X** of the expression is ((v1/v2) * v3)

and the **Y** of the expression is ((v4 – v5) * v8)

and if the expression 'X NE Y' (X not equal to Y) is *true* then the variable *index* is set at 0.

Parentheses (round brackets) are always required around the entire logical expression to be evaluated. The position of any inner parentheses within these external parentheses determines the nature of the logical expression by changing the order of evaluation. Thus in the command:

```
SPSS/PC: IF (((v4 + v5) * v6) GT 100) index = 3.
```

for each case, the value of *v4* is added to the value of *v5* and the sum then multiplied by the value of *v6*. If this product has a value greater than 100 then the expression is *true* for that case and the value of *index* is therefore set to 3. This sequence of operations is *not* the same as that requested by the command:

```
SPSS/PC: IF ((v4 + (v5 * v6)) GT 100) index = 3.
```

Here, for each case, the value of *v5* is multiplied by the value of *v6* and the product then added to the value of *v4*. If the resulting value is greater than 100 then the expression is *true* and the value 3 is assigned, for that case, to the variable *index*.

The logical expression to be evaluated as *true* or *not true* can be more complex in terms of its logic as well as its arithmetic. Several 'relations' (e.g. 'v1 GT v2'; 'score1 EQ score2') can be joined within a single **IF** command by the logical operators **AND**, **OR** and **NOT** to form a compound expression.

For example, the final evaluation of a student's performance might depend on both a continuous course-work component and the grade obtained in an end-of-course examination. An overall pass might require *both* that a score of 50 or more be obtained in the final exam *and* that the number of *c*ontinuous *a*ssessment *e*ssays *p*assed (the variable *caep*) be more than 5 (**GT** 5). The following expression would assign the value 1 for *pass* only to those candidates who were successful on *both* criteria:

```
SPSS/PC: IF (test GT 49 AND caep GT 5) pass = 1.
```

Another useful logical operator is **OR**. If, in a variation of the above example, a pass required *either* that a score of 50 or more be obtained in the final test *or* that more than 5 continuous assessment essays be passed ('*caep* **GT** 5') the following expression would assign the value 1 for *pass* to those cases which reached *either* of these criteria:

```
SPSS/PC: IF (test GT 49 OR caep GT 5) pass = 1.
```

A final variation of this example involves a complex situation. Suppose that a student is awarded a pass for a course if s/he has passed 5 continuous assessment essays and also passed the final examination (**GT** 50) *or* if 70 marks or more have been obtained in the examination alone. The following command could be used to identify which students had passed:

```
SPSS/PC:IF ((test GT 49 AND caep GT 5) or (test GE 70)) pass = 1.
```

The terms **AND** and **OR** may be used several times in a logical expression. Thus the expression:

```
(v1 ge 100 and v2 ge 100 and v3 ge 100 and v4 ge 100)
```

is true only if *all* of the four specified variables have values of 100 or more. By contrast, the expression:

```
(v1 ge 100 or v2 ge 100 or v3 ge 100 or v4 ge 100)
```

is true if *any* of the four variables has a value of 100 or more.

If two or more relational expressions are included within a single logical expression they must all be expressed in full form, as in the following examples:

```
IF (var001 LT 3 or var001 GT 5)
IF (v1 EQ 1 OR v2 EQ 1 OR v3 EQ 1)
```

The shortened forms 'IF (var001 LT 3 or GT 5)' and 'IF (v1 or v2 or v3 eq 1)' are *not* permitted.

Although each of the examples provided has involved the creation of a new variable (*pass* or *index*), existing variables can be recoded using the **IF** command. Thus, in the expression:

```
IF (v2 lt v3) v1 = 0
```

v1 could be a variable which already exists in the active file. For any case for which the logical expression was *true* the existing value of *v1* would be changed to 0. An extreme form of such a transformation can be seen in the command:

```
IF (v2 lt v3) v2 = 0
```

Here *v2* appears both within the logical expression and as the target transformation variable. The effect would be to recode *v2* to 0 for each case in which the initial value of *v2* was less than the value of *v3*.

One useful application of a conditional transformation of an existing variable involves using the variable as a counter. The value might be increased by one, for example, whenever some condition happens to be true. Thus the variable *agree* might be used to count the number of times someone has agreed with one of ten statements. Suppose that the variables related to the 10 statements named *state1* to *state10*, and that agreement is coded 1. The series of commands:

```
SPSS/PC: IF (state1  EQ 1) agree = agree + 1.
SPSS/PC: IF (state2  EQ 1) agree = agree + 1.
etc.
SPSS/PC: IF (state10 EQ 1) agree = agree + 1.
```

would eventually produce a value for *agree* equal to the number of times the respondent agreed with any of the 10 statements.

However, if *agree* did not exist as a variable *before* the first of the series of commands above, each case for which the first **IF** command was *false* would be assigned a system-missing value for *agree*. And once *agree* had been declared system-missing for a case, any transformation of *agree* would remain system-missing. Thus, following the above set of commands, *agree* would have a value equivalent to the number of statements agreed with only for those cases for which the logical expression in the first command was true (that is, assuming that *agree* was not previously established as a variable).

The way to avoid the difficulty is simply to initialize *agree* with a value of 0 before issuing the above series of demands. Thus we create a new variable, agree, and assign the value 0 for all cases, before including the variable in a conditional expression. This could be achieved by the simple **COMPUTE** command:

```
SPSS/PC: COMPUTE agree = 0.
```

This initialization command would assign the value 0 to the new variable *agree* for all cases. The first of the series of commands above would then increase the value of *agree* to 1 for those cases for which *state1* had the value

1. For other cases the value would remain at 0. The system-missing problem would have been overcome.

Where a subject has a missing value for a variable specified in the logical expression to be evaluated, SPSS/PC+ nevertheless attempts to evaluate the overall expression. For example, in:

```
SPSS/PC: IF (v1 EQ 1 OR v2 EQ 1 OR v3 EQ 1) score = 4.
```

a particular case may have a missing value for *v1*, but may have the value 1 for *v2* or *v3*. In such a situation the truth of the overall expression can be evaluated and the subject's *score* value *would* be set to 4.

Where the expression *cannot* be logically evaluated, however, the new value cannot be assigned and the new variable (*score* in the example we are now considering) is assigned a system-missing value. Thus if variables were joined by **AND**, rather than by **OR**:

```
SPSS/PC: IF (v1 EQ 1 AND v2 EQ 1 AND v3 EQ 1) score = 4.
```

then a subject with a missing value for *any* of the three variables (*v1*, *v2* **or** *v3*) would be assigned a system-missing value for the new variable *score*. If, however, *score* already exists as a variable in a dataset then, where the truth-value of the logical expression cannot be evaluated, the *original* value will be retained. With 'old' and 'new' codings likely to be intermingled, such a situation could easily lead to some confusion. The best strategy, for most purposes, is probably to use the **IF** command only to create *new* variables.

As with the other data transformations, the use of a conditional transformation (**IF**) command may require (*after* variables have been introduced or changed by the data transformation command) additions or revisions to variable labels and value labels.

When several data transformation commands are to be issued in a single interactive session, the order in which the commands will be entered should be carefully considered.

Thus if you intend to **RECODE** *v3* at some point, and you are also going to use *v3* in a **COMPUTE** command, you must decide whether you wish to **COMPUTE** the new variable using the original coded values for *v3* or using the *re*-coded values. Similarly, if you are going to evaluate a logical expression involving a *computed* variable you will obviously need to create the new variable *before* you refer to it in a conditional (**IF**) command.

8.4 Selecting cases

Sometimes an analysis needs to be performed on only *some* of the cases in the current active file. The cases to be selected for inclusion in an analysis may be:

- cases which have a particular characteristic (e.g. *sex* = male)
- the first *N* cases in the file (e.g. *N* = 100, or *N* = 300)
- a random sample (e.g. a random 1 in 10 cases)

Such a selection may be required to remain in effect for the rest of the current session, or for the execution of just one procedure.

Some of the selection commands discussed below will cause cases which are *not* selected to disappear from the active file. The only way of retrieving these 'lost' cases will be to **INCLUDE** the original file (e.g. "gender.def") once again. But in doing this, the effects of any transformations which have previously been made (e.g. with **COMPUTE** or **RECODE** commands) will be cancelled. In Chapter 9 we will consider how transformation commands can be inserted within the data definition file, to be executed whenever that file is **INCLUDED**. At this stage, if you have been following the text by issuing commands interactively (i.e. at the SPSS/PC: prompt), you must decide whether you really wish to remove cases 'permanently' from the current active file (i.e. until the end of the session). You may choose, instead, simply to read the instructions below, and to keep your full current file, including any transformations you have made. Some of the selection commands which follow, however, have only a 'temporary' effect and will not lead to cases being removed from the active file. You can explore the use of these temporary commands without affecting the contents of the active file.

To select cases with a particular characteristic

The **SELECT IF** command can be used to 'permanently' select cases. Thus:

```
SPSS/PC: SELECT IF (SEX EQ 1).
```

would be used to select only those cases in which the variable *sex* had the value 1. The truth-value of the logical expression would be evaluated for each case separately. Where the expression was 'true' the case would be selected *in*. Where the expression was 'not true' (or could not be evaluated) the case would not be selected and would be removed 'permanently' from the active file. Thus the above command would be used to select for subsequent analyses all the male cases, only, from a population in which males were coded 1 and females were coded 2.

As well as **EQ**, the other relational operators **NE**, **GT**, **GE**, **LT**, and **LE** can be used, as in the conditional transformation (**IF**) commands discussed above.

The **SELECT IF** command is also similar to the **IF** command in other ways. Complex conditional statements involving the logical operators **AND**, **OR**, and **NOT** can be used, as can numerical expressions, for example:

```
SPSS/PC: SELECT IF (WST GT 10 or HTF GT 5).
SPSS/PC: SELECT IF (var001 + var002 + var009 GT 16).
```

In this last example the values for the three variables for each case would be added together first, and the sum then evaluated in order to see whether, for that case, the sum was greater than 16. Those cases for which this was *true* would be selected, and cases for whom the expression was *not true* would not be selected. The reason that the three variables are added first is that arithmetic operations are evaluated *before* relations. The use of parentheses can make things clearer to the user, however, especially when the command includes a complex mix of arithmetic operations, relations and logical operations. The last example could be written:

```
SPSS/PC: SELECT IF ((var001 + var002 + var009) GT 16).
```

Brackets are also used to override the default order of evaluation of a logical expression and to specify a particular order to the system. For example:

```
SPSS/PC: SELECT IF ((100 - v1) ** 2) * (v2 - v3)  GT 100).
```

In the command:

```
SPSS/PC: SELECT IF ((var001 le var002) or (var003 gt var001)).
```

only the outer pair of parentheses are necessary. The extra parentheses change nothing, as far as the system is concerned, but may make clearer to the user what s/he is doing in issuing this command.

In this example, a case would be selected for analysis if, and only if, the value of *var001* was less than, or equal to, the value of *var002* **OR** if the value of *var003* was greater than the value of *var001* (or if *both* of these conditions were fulfilled).

A **SELECT IF** command affects all subsequent procedures (and thus has a 'permanent' effect). Examine the following example from the Gender Attitudes Survey:

```
(a) SPSS/PC: frequencies var001.
    SPSS/PC: SELECT IF (sex eq 1).
    SPSS/PC: frequencies var002.
    SPSS/PC: crosstabs var004 by var005.
```

Assuming that the active file initially includes all of the cases created by the **INCLUDE** "gender.def" command, the first procedure ('frequencies *var001*') would be performed on *all* 50 cases.

After the **SELECT IF** command, however, the subsequent procedures (both **FREQUENCIES**, specifying *var002*, and **CROSSTABS**) would be performed only with those cases selected to remain in the active file (in this example, only the 22 male cases).

Because only the male cases now remain in the active file, any subsequent procedure will also operate only on the 22 male cases. Thus any **RECODE** or **COMPUTE** command issued at this stage would affect only the male cases. To regain the full complement of 50 cases (to 'bring back the females') we would need to **INCLUDE** the original data definition file once more.

Because a **SELECT IF** command affects all subsequent procedures, multiple uses of the command will lead to a *progressive* selection. Thus, in the example:

(b)
```
SPSS/PC: SELECT IF (sex eq 1).
SPSS/PC: frequencies var002.
SPSS/PC: SELECT IF (var001 gt 3).
SPSS/PC: crosstabs var004 by var005.
```

the **CROSSTABS** procedure will be performed only on those cases which have the value for 1 for *sex* (specified in the first **SELECT IF** command – there are 22 such cases) *and* which have a value greater than 3 for *var001* (specified in the second **SELECT IF** command). The number of cases in the active file has been reduced to just 11.

It is possible to specify that the selection of cases should apply only to the procedure which immediately follows, i.e. that a temporary selection should be made. In such circumstances the **SELECT IF** command is replaced by **PROCESS IF**. Thus if we substitute **PROCESS IF** in the above example (a), giving:

(c)
```
SPSS/PC: frequencies var001.
SPSS/PC: PROCESS IF (sex eq 1).
SPSS/PC: frequencies var002.
SPSS/PC: crosstabs var004 by var005.
```

then (assuming that we started with the full complement of 50 cases) the first command ('frequencies *var001*') would operate on all cases. The procedure immediately following the **PROCESS IF** command (the **FREQUENCIES** command specifying the variable *var002*) would be performed using only the selected (22) cases. For any further procedures specified, however, *all* of the 50 cases in the original file would be included in the analysis. Thus the **CROSSTABS** procedure would be performed using *all* valid cases in the original population.

Thus with a series of suitable **PROCESS IF** commands, each followed by a particular procedure, different samples of the population could be analyzed successively. After each analysis the complete original set of cases would again be available.

Although the use of any of the relational operators (**EQ**, **GT**, etc.) is permitted within **PROCESS IF** commands, the logical operators (**AND**, **NOT** and **OR**) *cannot* be used.

To select the first N cases in a sample

The **SELECT IF** and **PROCESS IF** commands are used to select cases which conform to specified criteria. Sometimes, however, the user may wish merely to reduce the number of cases rather than to select cases which are of a particular type. One way of achieving such a size reduction is to specify the **number** (**N**) of cases to be selected. For example:

```
SPSS/PC: N 40.
```

This specifies that future procedures should operate only on the first 40 cases in the file. As with the other transformation commands, however, the **N** command will not in itself produce an immediate effect. The system will note the request but the selection will not affect the active file until the next procedure command is issued to the system.

The question of 'permanence' is rather complex in the case of **N** commands. The first **N** command creates a working dataset and, once this has affected the active file, the number of cases cannot be increased during the rest of the session. Thus the *first* **N** command reduces the number of cases in the active file and can therefore be regarded as having a permanent selection effect, while any further **N** commands issued during the session do not affect the number of cases in the active file.

For example, if we have used an initial **N** command to establish a working dataset of 40 cases, any further **N** command must specify a lower number. However, after a second **N** command has been issued to reduce the sample, a third **N** command can increase the size of the sample again, but only up to a maximum of 40. Thus a second **N** command '**N 20**' can be executed and then followed by a third command '**N 35**' (but *not* '**N 45**' because that exceeds the size of the working dataset created by the first **N** command). Thus any **N** commands issued after the first one may be regarded as having only a temporary effect.

The following sequence of commands may make this clearer:

Command	*Cases available for procedures*
include "gender.def".	(50)
N 40.	(50 – procedure not yet executed)
frequencies sex.	(40 – the first 40 cases of the original file, only, now form the working dataset)
N 20.	(40)
frequencies fac.	(40)
N 25.	(40)
frequencies yr.	(40)

There is, however, a further complication. If a transformation is requested (by, say, a **RECODE** command) and activated after **any N** command then the effect of that **N** command will be permanent. There is a good reason for this. If, having established a working dataset of 40 cases with an initial **N** command, a second **N** command '**N 30**' was followed by a **RECODE**, and '**N 35**' was then allowed to increase the number of cases, the first 30 cases processed subsequently would have the recoded values and the last 5 cases would have the original values. With some cases having recoded values and others having the original values, this would produce a confusing set of results in any subsequent analysis.

One last point. If an **N** command immediately follows, *or is immediately followed by* a **SELECT IF**, a **PROCESS IF** or a **SAMPLE** command (see below), then the **N** command is executed *after* that other command. Thus even if the command **SELECT IF** occurs after '**N 10**', the conditional selection is made first and *then* the first 10 **SELECT**ed cases are used for subsequent analyses.

To select a random sample

An alternative strategy for selection (which is always temporary and therefore always affects only the next occurring procedure) involves specifying a *proportion* of cases to be *randomly* selected from the total file.

This strategy involves using the **SAMPLE** command to select a random sample from the total population of cases in the active file. The command can be used in either of the following ways:

```
(a) SPSS/PC:    SAMPLE  .10.
(b) SPSS/PC:    SAMPLE  100 from 1010.
```

In example (a) the user has specified that a 1 in 10 random sample is to be taken. The system will select a random 10% of cases (approximately).

In example (b) the user has specified how many cases, **s**, (in this instance 100) are to be randomly sampled from the total file population (which, in this instance, the user is claiming to be, **S**, 1010). If a mistake is made in estimating the size of the total file population, adjustments are automatically made to the sampling. If the true population size is *less* than **S**, then the number sampled is proportionately less (thus if the true size were 909, the number sampled would be 90). If the true population size is *more* than that specified, then a full-sized sample (containing **s** cases) is taken, but that sample is taken randomly from only the first **S** cases.

It must be emphasized that the process of sampling, following a **SAMPLE** command, is a random one and that even if the same number (or proportion) of cases is sampled on a number of occasions the particular *cases* sampled will be different. In an exercise to explore the effect of this, file "gender.def" was made active and exactly the same pair of commands:

```
SPSS/PC:  sample 25 from 50.
SPSS/PC:  frequencies var001/ statistics = mean.
```

was issued 10 times. The mean values of *var001* for the 25 cases selected on each occasion were noted, and these ranged from 2.480 to 3.560 (the mean for all 49 valid cases in the file is 3.102).

8.5 Sorting cases

The order in which cases occur within the active file can be changed by a **SORT** command. This arranges cases in ascending or descending order of their value on any variable. **SORT** is a procedure command and therefore, unlike transformations commands, it produces an immediate response.

The default order for a **SORT** command is *ascending*. To reverse this we must specify a 'descending' sort (**D**). Thus, with the variable *age* coded in terms of years:

```
SPSS/PC: SORT by age.
```

will sort the cases in ascending age order (i.e. from the youngest to the oldest).

A descending sort would place the oldest 'above' the youngest:

```
SPSS/PC: SORT by age (D).
```

To observe the effect of a sort, the **LIST** command can be used. Thus if, after issuing the above command, **LIST** is requested, the values for all of the variables will be listed case by case. Using this example, it can be observed from the **LIST** display that the *id* variable is 'out of order' as the cases are displayed, but that cases are now presented in descending order of *age*.

If a **SORT** command has been issued *interactively* then, as long as the file remains active, the original order can be restored. If the file originally contained cases in the order of an *id* variable, of course, the initial order can be restored (for any cases which remain in the active file) by issuing the command:

```
SPSS/PC: sort by id.
```

In other situations, advantage can be taken of the fact that SPSS/PC+ assigns a system variable *$casenum* to each case (in the order it occurs within the file) as a file is made active. The original order can be restored at any time in the session by using *$casenum* as the sorting key:

```
SPSS/PC: sort by $casenum.
```

Sorts involving multiple criteria (or 'sort keys') can be specified. Thus:

```
SPSS/PC: SORT by age fac (D).
```

would produce a sort in which cases were arranged first in descending order of age. If two or more cases shared the same age, then the next sorting criterion *fac*, the respondent's faculty, would be used to order cases *within the same age group*. Again the order would be in terms of descending value (in this case 'social studies', coded 3, then 'science', coded 2, and then 'arts', coded 1). Notice from this example that the single **D** at the end of the command suffices to specify a descending sort for *both* of the variables *age* and *fac*.

An ascending sort can be specified by using **A**, so that the sorting command can specify any mixture of ascending and descending sorts. Thus, resulting from the following command:

```
SPSS/PC: SORT by fac (A) age (D) salary (A).
```

the initial, or major, sort would be an *ascending* sort in terms of *fac* ('arts', then 'science', then 'social studies'). Cases within the same faculty would then be ordered by descending age. Cases with both the same *fac* coding and the same *age* would be further sorted by ascending *salary*.

Precisely the same order of cases could be achieved by issuing three separate **SORT** statements but special care must be taken with the order in which they are issued. The major sort must come *last*, so the three **SORT** commands used to achieve the same result would be (in this order):

```
SPSS/PC: SORT by salary (a)
SPSS/PC: SORT by age (d)
SPSS/PC: SORT by fac (a)
```

This illustrates the fact that any **SORT** starts with the cases in their current order (not in the original order) and does the minimal amount of re-ordering required to satisfy the **SORT** request.

Combining selections and sorts

Sorting and selection commands can be combined to produce many different kinds of samples for analysis. The following examples illustrate some effects of combining such commands. In each case the commands specified were issued immediately following the inclusion of the *full* "gender.def" file (i.e. **50** cases).

```
(a)     SPSS/PC: sort by sex (D).
        SPSS/PC: sample 14 from 28.
  (i)   SPSS/PC: frequencies var001.
  (ii)  SPSS/PC: frequencies var002.
```

Since the variable *sex* is coded 1 for males and 2 for females, the *descending* sort of cases by *sex* places the 28 females in the total sample at the beginning of the active file. The temporary **SAMPLE** command samples 14 from the first 28 cases (here, all of the female cases) for analysis (i). All 50 cases are analyzed by command (ii).

```
(b)     SPSS/PC: select if (sex EQ 2).
        SPSS/PC: sample 14 from 28.
  (i)   SPSS/PC: frequencies var001.
  (ii)  SPSS/PC: frequencies var002.
```

Here the **SELECT** command remains in effect for all subsequent procedures whereas the **SAMPLE** command is temporary. Analysis (i) therefore involves 14 (female) cases, and analysis (ii) involves all 28 females (only – the males have been 'permanently' excluded by the **SELECT IF** command).

```
(c)     SPSS/PC: process if (sex EQ 2).
        SPSS/PC: sample .5.
  (i)   SPSS/PC: frequencies var001.
  (ii)  SPSS/PC: frequencies var002.
```

Here the **PROCESS IF** command leads to the temporary selection of the 28 female cases. **SAMPLE** leads to the inclusion of approximately 50% of these. Thus about 14 cases will be analyzed as a result of command (i). At stage (ii), however, neither of the temporary selections is operative, and data from all 50 cases are included in the analysis.

In this chapter we have considered how values may be recoded and how new variables may be created, and we have examined the ways in which cases may be selected, sampled and sorted.

9 *Command Files and System Files*

By now you should be familiar with the use of some or all of the procedures **FREQUENCY**, **CROSSTABS**, **MEANS** and **T-TEST**. You should also have explored the ways in which variables can be recoded, and new variables computed, and the manner in which cases can be selected or re-ordered.

In this Chapter we consider two types of file – **command files** and **system files**. We already have some experience of one kind of command file – the data definition file. System files, however, have not yet been considered.

Let us first recall the story so far. The editor program **REVIEW** has been used to create data files and data definition files for use by SPSS/PC+. Thus:

- Data have been entered and the dataset then written to disk as a **data file**.
- A **data definition file** has then been written. This lists the variables represented within a specified data file, with formatting information, variable labels and value labels. Missing values are assigned to variables and the complete data definition file then written to disk.

After a data definition file has been written to the disk it has been named in an **INCLUDE** command, such as

```
SPSS/PC: include "gender.def".
```

and an initial procedure command such as **LIST** OR **FREQUENCIES** then has the effect of creating an **active file** within the computer memory (unlike the other files mentioned, an active file is not a file which is stored on disk).

Once the active file has been created, the data within it can be processed and analyzed in various ways. Commands and subcommands can then be issued interactively (i.e. in response to the SPSS/PC: prompt or the continuation prompt :).

This is a good time to recapitulate on the various types of SPSS/PC+ command.

9.1 Types of command

Commands can be divided into three different groups:

(a) **operation commands** provide information or change parameters of the SPSS/PC+ operation – they do not transform or analyze data. This group of commands includes **DISPLAY**, **SHOW**, **SET**, **REVIEW**, **INCLUDE**, and **FINISH**. They are executed immediately after the command is issued.

(b) **data definition and manipulation commands** provide information to the system about the dataset or about how it should later be used. Such commands may specify the file in which data are to be found, and describe how the dataset is arranged; they may provide information about variable and value labels, transformations, and the selection of cases. So far we have met **DATA LIST**, **VARIABLE LABELS**, **VALUE LABELS**, **MISSING VALUES**, **RECODE**, **COMPUTE**, **SELECT IF**, **PROCESS IF**, **N**, and **SAMPLE**. These commands are not executed immediately but are noted by the system and executed when a subsequent procedure command is entered.

(c) **procedure commands** read the data and, usually, perform analyses. The initial procedure command (for example, **FREQUENCIES**) following a 'read data' command (for example, **DATA LIST**) creates an active file as defined by the 'read data' command. Subsequent procedure commands read the data from the current active file. When such a procedure command is entered, any data definition or manipulation commands which have been noted by the system but not yet put into effect are first executed, and the procedure itself is then executed. We have so far encountered the procedure commands **LIST** (which displays data values) and **SORT** (which re-orders cases in the file) neither of which analyze data, and the statistical procedure commands **FREQUENCIES**, **CROSSTABS**, **MEANS**, and **T-TEST**.

So far most of these commands have been entered interactively, although some have been written into the data definition file (**DATA LIST**, **VARIABLE LABELS**, **VALUE LABELS**, **MISSING VALUES**). The commands which have been entered interactively, however, can also be written into files. These are known as **command files** (and thus a data definition file is a special type of command file).

9.2 Command files

When an interactive session is brought to an end (with the command **FINISH**), the original data definition file remains unaltered on the disk, ready to be retrieved at a later date for further analysis. However, any transformations (including re-codes and new computed variables) are lost and the relevant commands will need to be entered again if the transformations are to be re-introduced.

An active file may continue to be worked on interactively, no matter how complex the transformations and analyses might become, but such a strategy is by no means efficient. The solution to this problem is to create a command file which includes such transformation (and perhaps other) commands. These commands are then executed by the system, one by one, as if they had been entered interactively from the keyboard. Thus a whole series of operation, transformation and procedure commands can be written as a seperate file, or added into an existing data definition file. When this file is **INCLUDED**, the sequence of commands is then executed automatically by the system. Processing controlled by a command file, rather than commands entered interactively in response to the SPSS/PC: prompt, is referred to as 'batch processing'.

The user may therefore either add further commands to the data definition file or create a new command file (a 'batch file').

Like the data files and data definition files we have already encountered, command files can be created or edited with an editor program. In the following discussion it will be assumed that the SPSS/PC+ editor **REVIEW** is used, although alternative editors and word-processors may also be used (see *Appendix B*).

To illustrate how a complex command file might be created from an existing data definition file, we can add some **RECODE** commands and some **COMPUTE** commands to the data definition file "gender.def" (from the Gender Attitudes Survey). In Chapter 8 some suggestions were provided for transformations which might be performed on data from the Survey. In particular, suggestions were given for:

- recoding *salary* into just three values (p. 103)
- computing the variables *wt* and *ht* (p. 104)
- scoring a test for *extra* and *neur* (pp. 106–107)
- computing a *sexism* score (pp. 108–109)
- computing a *problem* score (p. 109)

If the relevant commands were issued interactively, at the end of the session the re-coded variables and the new computed variables would be lost. Such transformations could be made a permanent feature of the file, however, by adding the relevant transformation commands directly into the data definition file.

To do this, we would first use **REVIEW** to bring the original data definition file forward for editing:

```
SPSS/PC: review "gender.def".
```

A copy of the contents of this file would now be displayed on the screen, with the cursor at the beginning of the last line. Moving the cursor to the end of this line and pressing <RETURN> would provide a new blank line. To request the re-codes and the new computed variables in the list above we could then add the following lines:

```
recode salary (LO thru 5000=1) (5001 thru 7500=2) (7501 thru 99990=3).
compute wt= (wst * 14) + wlb.
compute ht= (htf * 12) + hti.
recode E1 to E12 (3= -1) (2= 0).
compute extra = (E2 + E4 + E7 + E8 + E10 + E12).
compute neur = (E1 + E3 + E5 + E6 + E9 + E11).
compute sexism = (var001 + var002 + var009 + (8 - var003) +
    (8 - var004) + (8 - var005) + (8 - var008)).
compute sexism = sexism - 7.
compute problem = (2 - prs01) + (2 - prs03) + (2 - prs04) +
    (2 - prs07) + (2 - prs08) + (2 - prs09) + (2 - prs10).
```

To bring the labelling up to date and in line with these transformations, we should also add the following label revisions:

```
variable labels salary "Recoded salary"
/ wt "Weight in pounds" / ht "Height in inches"
/ extra "Extraversion score" / neur "Neuroticism score"
/ sexism "Sexism score 0-42 from attitude variables"
/ problem "Problem score 0-7 from prs variables".
Value labels salary 1 "LO thru 5000" 2 "5001 thru 7500"
3 "7501 thru 99990 (missing NOT included)"
/ E1 to E12 1 "yes" 0 "not sure" -1 "no".
```

Procedure commands could also be added into the file. For example, we could request **FREQUENCIES** for some of the newly created variables by adding:

```
frequencies extra neur problem sexism
/barchart
/statistics = mean stddev mode range.
```

The edited file could then be written to disk by using the **REVIEW** command **<F9><w>**. We would be prompted with the original file name ("gender.def") and if we wished to accept this name for the enhanced file we would simply press <RETURN>. However, we might wish to preserve a copy of our original data definition file, without the transformation commands, etc., and we could therefore supply an alternative name. Command files are often given the extension ('surname') .SPS, and so we might decide to use the name "gender.sps" for the enhanced file.

The command file "gender.sps" will now be available on the disk, and because a new name has been specified for this file, it will not have overwritten the original data definition file. Thus "gender.def" will still be available.

To execute the series of commands contained within the command file we would first exit from **REVIEW** (using **<Alt-F10>**) and then respond to the SPSS/PC+ prompt with an **INCLUDE** command specifying the new command file. For example:

```
SPSS/PC: include "gender.sps".
```

As the command file is being **INCLUDED**, the transformation commands will be noted by the system. These will be executed immediately before a subsequent procedure, and any procedures specified within the command file will then be performed.

If any problem emerges during the inclusion process, an error message will be displayed. The command file can then be corrected by issuing a **REVIEW** command such as

```
SPSS/PC: review "gender.sps".
```

and editing out any errors in the normal way.

When a command file has been successfully included, and after any transformations and specified procedures have been performed, the file will remain as the active file and the SPSS/PC: prompt will be presented.

If further analyses are required on the same dataset there are several ways of proceeding:

- the current active file (which contains transformations, etc.) can be processed interactively
- the existing command file (based on the original data definition file) can be re-edited with **REVIEW**
- a new additional command file can be created with **REVIEW**.

If the existing command file is to be revised (the second possibility listed above), then some of the existing procedure commands may need to be deleted (using the line delete function **<F4><d>**), to prevent the system from repeating analyses which have already been executed. New procedure commands could then be added. When the revised command file has been written to disk, and an exit made from **REVIEW**, the revised command file would need to be **INCLUDED** in response to the SPSS/PC: prompt.

Alternatively, a new command file can be written. Such a file may be prepared in one session for use in a later SPSS/PC+ session, or may be written while the current active file is still available (since using **REVIEW** does not affect the active file). If this is done, then when the new command file is **INCLUDED** the commands it contains will operate immediately on the active file.

If a new command file is to operate on a dataset which is not currently the basis of the active file, an appropriate **INCLUDE** command can be placed at the beginning. When the command file is then executed the relevant data definition file will be included before subsequent commands are issued from within the command file.

Thus, the following command file merely 'refers' to the relevant data definition file with an initial **INCLUDE** command:

```
include "gender.def".
recode E1 to E12 (2= 0) (3 = -1).
compute extra = (E2 + E4 + E7 + E8 + E10 + E12).
compute neur = (E1 + E3 + E5 + E6 + E9 + E11).
frequencies extra neur
/statistics = mean stddev mode range.
```

This is an example of a complete command file. It could be written to disk as, say, "extra.sps". At any later stage, the interactive command:

```
SPSS/PC: include "extra.sps".
```

would be sufficient to produce a display of the frequencies requested by the command file.

The first line of this command file will lead to the inclusion of "gender.def", which in turn contains a reference to the relevant **data file** ("gender.dat"). The active file will therefore include information which has been obtained from three relevant files although only one of these ("extra.sps") has been directly specified in the interactive **INCLUDE** command.

Command files are especially useful when a similar sequence of procedures is to be performed a number of times (e.g. with different variables, different **OPTIONS**, or different **STATISTICS**). Rather than issuing a series of similar commands repeatedly in the interactive mode, a command file could be written once and then edited a number of times to produce the desired variations on the initial analysis.

Operation commands (**SET**, etc.) can also be added into command files. Thus executing the command file:

```
include "gender.def".
set printer on.
frequencies var001.
set printer off.
frequencies var002.
```

will produce a screen display of both **FREQUENCIES** commands. Only for the variable specified on the first of these commands (i.e. *var001*), however, will there be a printed record.

Command files which contain a different batch of useful operation commands can be created. Given different names they will remain in the current directory until deleted, and since they make no reference to any specific file they will be available as general tools when you are working with any active file.

If such a file of operation commands is created and written to the disk with the name "SPSSPROF.INI" it will be automatically executed each time that SPSS/PC+ is run from the current directory. If this is your own directory, you will be able to 'profile' your own tailored way of working with SPSS/PC+. Thus if you were working continually on the "gender.sps" file, and preferred to work with the beep switched *off* and the printer switched *on*, the following "SPSSPROF.INI" file would set a useful profile:

```
set printer on.
set beep off.
include "gender.sps".
```

Your own SPSSPROF.INI commands will be executed when *you* use SPSS/PC+ from your own directory (automatically via a pre-set path command in the **DOS** 'autoexec.bat' file). When you enter SPSS/PC+ the SPSSPROF.INI commands will be displayed, immediately after the logo display, and an 'End of Profile' message will be given before the system presents the opening display. Other people, who enter SPSS/PC+ from their own directories, may work with a different profile set by their own quite different SPSSPROF.INI file.

As your project reaches a new phase, or as you change from one project to another, you will want to change the commands within this file. You can delete the file at any time (from **DOS**, using the 'del' command), you can edit the existing "SPSSPROF.INI" file with **REVIEW**, or you can use **REVIEW** to create another file of the same name (and this new file, when saved, will automatically overwrite the old version).

9.3 Using the log file

The above discussion of batch files omitted to mention one further way of creating such files. As well as creating such files afresh (starting with **REVIEW** and a blank screen, and then writing a list of commands) or re-editing an existing command file, it is possible to create a batch file using commands which have previously been entered interactively (i.e. in response to the SPSS/PC: prompt).

This takes advantage of the fact that SPSS/PC+ automatically keeps a record of all commands issued during a session (whether they are entered interactively or in a batch file). Thus if the user has been issuing commands interactively, and then needs to repeat certain analyses (perhaps with different variables, or with different subcommand options) this can be achieved by editing a record of previous interactive commands.

Just as SPSS/PC+ automatically creates a **listing file** (with the default name "SPSS.LIS") containing all of the output for a session, it also creates a file containing a record of all commands issued to the system. This is the **log file** (akin to a ship's log) and is named (by default) "SPSS.LOG". When a session begins, the existing "SPSS.LOG" file is initialized (and the contents therefore erased). Any command entered during the session is recorded as the session proceeds, and this provides an editable record which can be useful in monitoring progress and detecting errors in commands. The log file contains a record of all of the commands entered during the current interactive session with SPSS/PC+, together with information that a warning or error was issued, and references to output page numbers.

For the log file, as for the listing file, a name other than the default name can be specified at the beginning of the session with a **SET** command. For example:

```
SPSS/PC : set log "gender.log".
```

The use of such a **SET** command will prevent the log file from the last session being automatically overwritten when a new session begins, although different names must be set for the log file each time if a series of successive records is to be maintained (because any file subsequently saved will lead to the original file with that name being erased).

If the name for the log file is **SET during** a session then the named file will contain the log *from that time on*. A record of the earlier part of the session will be retained in a separate file with the original name.

We can use **REVIEW** within a session to edit an appropriate section of the log file and then to write this as a command file which can be **INCLUDED** to bring about a batch processing of the current active file.

Like listing files, log files are stored in ASCII form and can be edited and printed using text editor and word-processor programs. (General information about using other editors or word-processing systems is provided in *Appendix B*). However, while listing files are more likely to be used *after* a session, to make a copy of the output for editing and printing, log files are more likely to be used *within* a session.

Either of the commands:

```
SPSS/PC :  review log.
```

or

```
SPSS/PC :  review "spss.log".
```

will produce a display of the most recent part of the log file (the second example assumes that the default log file name has been retained).

The normal editing commands introduced in Chapter 4 are used (and you might well make use of some of the more advanced **REVIEW** editing facilities to be introduced in the next chapter). However, there is one special point that needs to be made with regard to the editing of log files, and if you are to save only a part of the edited log file as a new command file, then you need to know how a 'block' within the file being edited (rather than the *whole* file) can be written to the disk as a separate file. We will deal with these two issues.

Dealing with square bracket lines

The special editing point relates to the square brackets ([) which usually precede many of the lines of text within a log file. Such lines of text may contain records of error messages, information which has been read into SPSS/PC+ via an **INCLUDE** command, or output page numbers. If the log file is being edited in order to create a new command file then special attention should be paid to any line with a square bracket.

If such a line is to be *active* element in the new command file the bracket must be deleted, because lines starting with such a bracket are not executed. Lines which refer to previous errors or to page numbers can be safely retained within the file, but such lines must *retain* the initial bracket if errors are not to occur. As long as the bracket remains at the beginning of each of these lines they will have no effect. They will not disturb the system or produce errors when the file is subsequently **INCLUDED**.

To save a 'block'

Typically, when using a log file as the basis for creating a new command file, we are concerned only with one or two 'relevant' sections of the log file (those which contain material similar to that which we wish to include in a new batch file). Thus we will be working only with a part of the log file, and when the time comes to save our work, we will wish to save only the relevant lines of information. We need, therefore, to make a 'block' within the file, and then to save the block. We would not wish to save a copy of the whole (edited) log file as a new command file.

In Chapter 4 we discussed the ways in which **REVIEW** allows us to mark areas of a file (blocks of whole lines, or rectangles) and then to copy, move or delete them using various commands that involve the function keys F7 and F8.

For present purposes we need only be concerned with 'blocks' of whole lines. To first mark a block of lines we place the cursor on the first line of the intended block, and then issue the command **<F7><l>**. The line will begin to flash and a message: **"Waiting for second block marker"** will appear. The cursor should now be moved to the line which is to be the last line of the block, and **<F7>** pressed again. The block is then specified, and the lines in the block will change to a different shade (or colour). A message confirms that the block has been marked: **"Area marked *nn* lines"**.

If you wish to cancel the block, simply press **<F7>** once. The highlighting will disappear and an **"Area cancelled"** message will be displayed. A marked area can be deleted by pressing **<F8><d>**, or copied to another part of the current file by moving the cursor to the position where the copy is to be inserted and entering **<F8><c>**. The block can also be moved from the current position to another by moving the cursor to the new position and entering **<F8><m>**.

Our concern here, however, is to save the block to a new file. You may recall that to write a *whole* file (i.e. to save it on the disk) we used the command **<F9><w>**. To save a *marked area* we use **<F9><m>**. The default name "REVIEW.TMP" will appear. It is unlikely that you will wish to accept this name (should you wish to do so, you need only press <RETURN>), so enter a name of your choice. The forename should be something unique, informative, and specific to your study, and you may choose to use as the file extension ('surname') ".SPS" (a common extension given to command files). When you have entered a name, press <RETURN> and the block will be saved as a separate file with the name you have specified. If you now wish to execute the batch of commands in this file, using the data within the current active file, exit from **REVIEW** (using **<Alt-F10>**) and **INCLUDE** the new command file in response to the SPSS/PC: prompt. For example:

```
SPSS/PC: include "bloksave.sps".
```

Each of the commands within this newly created batch file will now be executed using the contents of the current active file.

9.4 System files

Another type of file which may increase the efficiency of operating with SPSS/PC+ is the **system file**. This is a complete record of the active file in its current state (i.e. with any selections, transformations, re-sorts, labels, etc. remaining in effect).

Within a system file, the information is stored in a special SPSS/PC+ format and written in binary form (i.e. as 'bits' of information) rather than in ASCII form (i.e. as alphanumeric characters). The advantage of working with a system file is that processing time is significantly reduced. Also, for some procedures, such as **JOIN** (to be discussed in Chapter 15) it is essential to work with system files. However, there is a major disadvantage involved in working with such files, because *system files cannot be edited.*

The commands for writing a system file to disk, and for later making such a file active, are not the same as those used for command files. System files are written to disk with a **SAVE** command and are subsequently read with a **GET** command (*not* with an **INCLUDE** command).

Thus, to save the current active file as a system file, we would issue a command such as

```
SPSS/PC: save outfile= "gender.sys".
```

If the **SAVE** command is used without specifying a name, the default name "SPSS.SYS" is used. Thus:

```
SPSS/PC: save.
```

will produce a system file named "SPSS.SYS". If a file with that name already exists it will be overwritten. **SAVE** copies the current active file but does not affect it; it remains as the active file until another file is made active or the session finishes.

A system file contains the current contents of the active file, including data and data dictionary information. Thus variables which have been recoded, for example, will be represented in terms of their new values (only). If a 'permanent' selection command has been issued, only the remaining cases will be represented in the system file. Thus, for example, if an active file had been made from the complete "gender.def" file, and the female cases then selected for analysis (with **SELECT IF**), only the data from the female respondents would remain in the active file. A **SAVE** command issued at this stage would therefore lead to the creation of a system file containing only the information relating to the female respondents. Similarly, if a **SORT** command has been issued, then when the file is **SAVED** the cases will be stored in their re-ordered sequence.

Thus special care is needed when working with an active file which is subsequently to be **SAVED**. The commands **RECODE**, **COMPUTE**, **IF**, **SELECT IF**, **N**, and **SORT** have a 'permanent' effect on the contents of the active file (and will thus affect the contents of any system file subsequently created from it).

Selection commands which have only a temporary effect, however (**PROCESS IF** and **SAMPLE**), do not change the active file and therefore have no effect on any subsequently created system file.

It is possible to issue a **Drop** subcommand within the **SAVE** command to request that not all of the variables in the current active file be included in the system file which is to be written to disk. The **Drop** subcommand should list those variables which are *not* to be included in the system file. The **TO** convention can be used, as in:

```
SPSS/PC: save outfile= "gender.sys" / drop fac yr prs01 to prs10.
```

The system file will now contain all of the variables in the current active file except those named on the **Drop** subcommand. But the 'dropped' variables will remain in the active file, because the active file is not affected when a **SAVE** command is issued.

In addition to the user-supplied information (data, labels, missing values, etc.) the system file will also contain extra information supplied by the system – **system variables**. One of these, *$casenum* was introduced when we considered the use of the **SORT** command in Chapter 8. *$casenum* is an identifier number given to cases (in the order in which they occur within the dataset) as they are read into the active file. Even after **SORT** commands have been issued, the *$casenum* for each case remains the same. The value for *$casenum* is stored with the other data for a case when it is **SAVED** as part of a system file. Thus it will be possible (by using a 'sort by *$casenum*' command) to re-order cases which have been read from a system file so that they are arranged in the same order as when the source file was first made active.

Once a system file has been written it can be read. Because of the special form in which such files are written, however, the **INCLUDE** command *cannot* be used to read system files into SPSS/PC+. Having been created using a **SAVE** command, system files must be read into SPSS/PC+ with a **GET** command.

GET, without a specified filename, will read the default system file "SPSS.SYS"). With a file name specified it will read that system file:

```
SPSS/PC: get file= "gender.sys".
```

This would lead to the dictionary of the system file "gender.sys" being read (**GET** is not a procedure command – the data would be read only when a procedure is subsequently issued).

A **GET** command can also be issued from within a command file. Thus if the system file "gender.sys" has previously been **SAVED**, we can **GET** it with the command file:

```
get file= "gender.sys".
frequencies var001 to var003
/statistics= mean stddev mode range.
```

As with **SAVE**, a **Drop** subcommand can be used with **GET**. Thus:

```
SPSS/PC: get file= "gender.sys" / drop var001 to var010.
```

The addition of a **Drop** subcommand has the added effect of creating an active file. If no **Drop** subcommand is used, then no active file is created until after a transformation or selection command has been issued. If only procedure commands are issued (and if no **Drop** subcommand has been included in the **GET** command) then the system reads the data from the system file each time it executes a procedure.

When a system file has been read it can be worked with interactively or through batch processing, so that by recoding and computing variables, changing labels, and sorting and selecting cases, a limited revision of the contents of the file can be achieved.

Provided that an active file has been created, the **SAVE** command can then be used to write a new system file. This might retain the original name or, alternatively, after the active file has been tailored interactively, it could be **SAVED** using a *new* name:

```
SPSS/PC: SAVE outfile= "gentrans.sys".
```

In this case (if "gentrans.sys" were NOT the name of the original system file) there would now be *two* relevant system files on the disk: the original (*without* the recent changes) and a new one, "gentrans.sys" (*with* the recent changes). Either of these could be called up in future with the appropriate **GET** command, for example:

```
SPSS/PC: GET file= "gentrans.sys".
```

or

```
SPSS/PC: GET file= "gender.sys".
```

Sometimes the user may wish to keep several different versions of the same file on disk, each with its different selection of cases, transformations, labels, etc. One or another of these can then be made the active file (with **GET**) as it is required in various stages of the project analysis.

It is clearly important to give meaningful names to files and to make a note of the state of the data in any particular file. A reminder of the names and labels of the variables in the file can be obtained by issuing the **DISPLAY** command. Fuller information can be obtained by naming particular variables on the **DISPLAY** command or by using **DISPLAY ALL**.

Whatever system files may be created, it is essential to retain on the disk the full details of the data in the form of the data file and the data definition file. As long as these files are not deleted, a fully editable version of the data and data dictionary will be available.

Indeed, although procedures will be more quickly and efficiently executed using system files, the user may prefer to work with the types of file which may be slower in operation but which have the distinct advantage of being editable.

Thus, instead of saving several different system files for the same study, a number of command files can be written, each specifying a different set of selections, recodes, transformations and procedures to be used on the dataset represented within a data file, and defined and labelled by the relevant data definition file.

10 Further Explorations of REVIEW

It is now time to extend our use of the range of facilities offered by the editor **REVIEW**.

Various **REVIEW** editing commands have been introduced in previous chapters, and the next section presents a brief revision of material already introduced. We will then consider some new editing features of **REVIEW** and some of the extended SPSS/PC+ facilities available from within **REVIEW**.

10.1 The story so far

REVIEW is the SPSS/PC+ editor program. To create a new file, or to edit an existing file, you issue a **REVIEW** command:

```
SPSS/PC: review "newfile.def".
```

You are presented with a blank screen on which you type the information you wish to have included in the file. Remember that there are two modes: 'insert' (the **Ins** message will appear at the bottom of the screen) and overwrite (or 'insert off'). Various special editing functions are implemented by pressing one or two keys, either together (e.g. **<Alt-F10>**, to exit from **REVIEW**), or sequentially (e.g. **<F4><d>**, to delete the current line). A new blank line can be *inserted after* a current line (i.e. the line with the cursor on it) by pressing **<F4><i>** or can be inserted *before* the current line by pressing **<F4><b>**. The last deleted line can be *undeleted* (at the original position or a new position) by the command **<F4><u>**.

It is possible to mark a specific area within the file being created (blocks of whole line, or rectangles), and marked areas can be deleted, copied, or moved by using various commands that involve the function keys F7 and F8 (see p. 54 for a list of these commands). A marked block can also be written as a separate file on the disk by issuing the command **<F9><m>**, and the whole file can be written to disk with **<F9><w>**.

In addition to these basic commands, there are various other functions which have not yet been introduced, including obtaining on-line help with the **REVIEW** commands, quick ways of searching through a file, efficient ways of changing 'strings' of characters within the file, and entering a different file into **REVIEW**. In this chapter we will also consider how on-line help with the contents of the commands you are writing can be obtained, and how an on-line glossary can be consulted from within **REVIEW**.

10.2 REVIEW help

If you find it difficult to remember the various commands used to bring about **REVIEW** editing functions, don't worry. The command **<F1><r>** produces a display of the **REVIEW** Help screen, showing the various uses of the function keys. When you know which 'F' key to press, return to editing (by pressing <SPACE> or **<Esc>**) and then press the appropriate function key. A minimenu will be displayed, and this will indicate which further letter-key you need to press to bring about the desired effect (or, alternatively, you can simply move the highlight, with the cursor keys, and press <RETURN>).

When you have pressed **<F1><r>**, and are presented with the display showing the uses of the function keys, pressing **<F1>** again will produce a display that explains the commands that bring about movement through a file.

The **<F1>** command has a different effect when the system is presenting a prompt (e.g. when it is asking for a filename to be specified). **<F1>** then causes the system to provide a short message explaining how you should reply to the prompt.

10.3 Paging and searching

Although it is possible to move within the file using just the cursor, there are more efficient ways of moving around within long log or listing files. Thus we can search for a particular set of characters (a 'string' – for example the name of a certain variable) or we can direct the cursor to a particular SPSS/PC+ page. Paging the listing (or output) file will be especially easy if we have a copy of the printed output, since the SPSS/PC+ page numbers on the file will be the same as the page numbers on the printout.

Finding a page in a listing file

To find a particular numbered SPSS/PC+ *O*utput page in a listing file, press **<F6><o>** and reply to the message:

```
Page number :
```

with the number of the page you wish to find. The cursor will then be placed at the beginning of that page. This command can only be issued when you are editing a listing file.

Searching for strings

A 'string' is simply a sequence of characters (which may include symbols and spaces). You may move the cursor from the current position to the first occurrence of any specified string which is above or below the current cursor position.

To initiate a *F*orward search for a string, press **<F5><f>** and you will be presented with the message:

```
Search string :
```

The string is now entered. Upper-case and lower-case characters are treated as equivalent, so that specifying "Neur" will find "neur" and "NEUR". The computer will search forward through the file for the first occurrence of the specified string and the cursor will flash on the first character of that string.

To find a second occurrence of the same string, **<F5><f>** must be pressed again. Once a string has been entered it becomes the default string, so that any subsequent occurrence of the same string can be found merely by pressing **<F5><f>** followed by <RETURN>.

A *B*ackward search can be implemented by pressing **<F5><b>** and specifying the string to be located. The search will begin from the current cursor position and proceed backwards through the file.

Changing strings

One of the most useful features of **REVIEW**, especially when editing command files, is its ability to replace one sequence of characters (a string) with another.

The usefulness of this facility can be illustrated using the example of the data definition file written to analyze the data from the Gender Attitudes Survey (this data definition file is reproduced in *Appendix F*).

Two sets of variables are parallel. Respondents were asked whether each one of 10 statements was more likely to receive agreement from men or from women (these are the 'stereotyping' variables *st01* to *st10*), and they were later asked, of precisely the same statements, in the same order, whether the statements were true of themselves (variables *prs01* to *prs10*).

In writing the data definition file, the variable labels applicable to both sets of variables would clearly be similar. In fact, the same brief versions of the statements are given, but whereas for the 'st' variables the term '(M-W)' (for 'men and women') is appended, for the 'prs' variables the term '(SELF)' is appended. Thus the label for *st1* is "Depressed often (M-W)" and the label for *prs01* is "Depressed often (SELF)".

It would clearly be a tedious to write out the set of labels twice, and it is clear that effort would be saved by copying the block of lines specifying the 'st' labels and then manually editing the copy so that 'st' is replaced by 'prs' when specifying each variable name, and '(M-W)' is replaced by '(SELF)' when specifying the label.

However, the same effect could be achieved with even less effort, by using the **REVIEW** facility for changing strings. After copying the block of lines specifying the 'st' variable labels, we make all 20 revisions (2 X 10) by issuing just two commands. With the cursor on the first line of the copied block (i.e. below the 'st' lines which are to be remain unaltered) we could start by changing the variable name. To **change forwards** from the current cursor position, we would issue the f*O*rward change command **<F5><o>**. In response to the prompt:

```
Old string :
```

we would enter the string to be changed (i.e. in this case 'st'). The string must be entered in *exactly* the form required. Upper-case and lower-case characters are *not* equivalent in change operations, so that specifying "Neur" will *not* find "neur".

After entering the string and pressing <RETURN> the prompt:

```
New string :
```

will appear. You then enter the string to be substituted (in this case 'prs').

The system will search forward through the file for the first occurrence of the specified string and the cursor will flash on the first character of that string. When the first occurrence of the string is located the user is presented with a minimenu display listing a number of options as described in Table 10.1.

Table 10.1

C=chg&next A=chg all N=skip&nxt S=stop X=chg&stop
The appropriate key – C, A, N, S or X – must now be pressed.
C replaces the string & proceeds to next occurrence of old string
A replaces this and every (subsequent) occurrence of old string
N leaves this occurrence intact and proceeds to the next
S leaves this occurrence intact and stops the search
X replaces this occurrence and then stops the search

Following a **C** or **N** response, the same list of options will be presented again when the next occurrence of the old string is found.

In the example (where we wish to change *all* of the occurrences of 'st' to 'prs') we could reply to the option display by keying **A** (= change this and all subsequent occurrences). The changes will be made automatically throughout the remainder of the file, and the computer will then issue a message detailing the number of changes made (e.g. **"10 changes made"**).

If a b*A*ckward change is required, rather than a forward change, then the replacement procedure should be initiated with **<F5><a>** rather than with **<F5><o>**.

We now wish to change the variable labels for the *prs* variables from '(M-W)' to '(SELF)'. To do this, return the cursor to the line on which we started making changes, and enter another **<F1><o>** command. This we enter the 'old string' – '(M-W)' – and the new string – '(SELF)' – and in response to the minimenu we again enter **A**, for 'change all'. Thus with two simple operations we have supplied labels for 10 variables, without the need for tedious and repetitive typing.

It will be clear that the **REVIEW** facility for changing strings will also be very useful when entering a long series of similar commands into a batch file.

10.4 Obtaining a list of variables

A display of the names of each variable in the current active file can be obtained by issuing the command **<F1><v>**. This will produce a two-window display at the top of the screen. The upper window displays the names of the first 24 variables in the active file, including a number of variables (one of which is *$casenum*) that have been added by the system. The first variable name is highlighted, and the lower window provides more detailed information on this variable, including the variable label if one has been supplied, the missing value, the type of variable (e.g. numeric) and the width of the variable (e.g. 2 for '2 columns'). The display also indicates whether value labels have been provided, although the labels themselves are not included.

The cursor control keys can be used to move the highlight from one variable name to another in the upper window, and each time a new variable is highlighted the information in lower window will change. With the cursor on the bottom line, moving it down further will cause the display in the upper window to 'scroll', revealing more variable names.

You can select any number of variables from the menu (move the highlight to each one in turn and press <RETURN>), so that they are automatically added to a command you are writing, and the word 'TO' is included at the top of the variable list so that you can even use this to specify a list of variables.

To return to manual editing, press **<Esc>**.

10.5 Calling another file into REVIEW

From within **REVIEW**, the user can ask for a file on disk to be directly inserted for inspection and editing. This is done by pressing **<F3><i>** and replying to the prompt

```
File to insert :
```

with the name of the relevant file. Certain files, of course (like system files), are not editable.

If you decide, when working within **REVIEW**, that you wish to edit a different file, the appropriate command is **<F3><i>**.

10.6 Listing file names

To request a list of files, press **<F1><f>**. The system replies with a prompt:

```
File specification :
```

and supplies a default specification (*.*). If this is accepted (by simply pressing <RETURN>) this will supply a list of *all* of the files in the current directory (whatever their forename or surname). A window at the top of the screen will display as many files as possible (the list begins with the oldest files and ends with the most recent), and you can move around the list using the cursor control, **<Page Down>**, and **<Page Up>** keys. As you scroll the display, further file names that are initially below the display window will be revealed.

Instead of accepting the default specification, however, you may wish to specify another file listing. A complete list of files in another directory (for example, in a directory named "study1" on *Drive D* could be requested by:

```
d:*.*
```

All files in the root directory of a floppy disk on *Drive A* could be listed as a result of issuing the request:

```
a:*.*
```

If you wish to list only *one* of the files in a particular directory, you can specify a particular file name:

```
gender.dat
```

and information relating to this file (the size of the file, in bytes, and the date and time when it was created) will be displayed on a single line just below the window that lists the file name.

You can also request a display list for a particular group of files. An asterisk (*) acts as a wildcard. Thus specifying **"*.sps"** would produce a list of every file in the current directory with the file extension **sps**. Specifying "**gender.***" would produce a list of all **gender**, files whatever their file extension. A **?** symbol acts as a wild character. Thus "**gend????.***" would produce a list of any file with **gend** as the first four letters (with or without any other characters), and with any file extension. The files "**gender.sps**" and "**gendatt.mat**" would be included in the list requested by this specification. If, with any wild characters, the length of the file name (without extension) is less than eight characters, then only files with names up to and including the specified length will appear in the list. Thus "**gend??.***" would produce a list which includes "**gender.sps**" but *not* "**gendatt.mat**".

When a list of files is presented in the window, use the cursor control keys to move the highlight to any one, and the information about file size and creation date will be presented on the display line below the file-name window.

When you wish to remove the display of file names, press **<Esc>**.

10.7 Using the 'help' facility

When editing command files, especially, it is often useful to be reminded of the alternative subcommands available for particular procedures, and the options available for any subcommand that you choose to use. Thus when using **FREQUENCIES**, you might wish to be reminded of the **Statistics** which can be requested and how these should be specified.

On-line information on such matters is accessible from within **REVIEW**. For example, when writing a **Statistics** subcommand for a **FREQUENCIES** procedure, you can request an on-screen reminder of the statistics available. To do this you would place the cursor on or just after the word "statistics" which forms part of the **FREQUENCIES** command that you are currently writing, and press **<F1><m>** (for *M*enu). An information window will appear at the top of the screen, displaying the options available within the **FREQUENCIES Statistics** subcommand.

Initially, the word 'DEFAULT' is highlighted in the left-hand box of the window, and the information in the right-hand box tells us that the additional statistics displayed when 'DEFAULT' is specified on a **FREQUENCIES Statistics** subcommand are the mean, the standard deviation, the minimum value and the maximum value.

If the highlight is now moved down two places, by pressing the 'down' cursor control key twice, 'MEAN' is highlighted in the left-hand box, and the right-hand box provides us with a definition and description of this statistic. Moving the highlight further down eventually causes the display to scroll, because the list extends beyond the initial window. You can also move to a new 'page' of the list using the **<Pg Dn>** and **<Pg Up>** keys.

To include any of the listed statistics as part of your command, move the highlight to the statistic you wish to choose and press <RETURN>. Do this for as many statistics as you want added to your list. Each time you press <RETURN> you will notice that the appropriate specification for the statistic you have chosen is automatically added to your command. If you wish to stop using the menu system and to return to manual editing at any stage, you do this by entering **<Alt-M>**.

The menu system is rather complex and is arranged in many different 'layers'. The **FREQUENCIES Statistics** example we have been working with, for instance, has been operating at the fifth layer deep in the menu system. To ascend the menu system by one layer, press **<Esc>**. This will present you with a range of alternative options for the **FREQUENCIES** command. You may wish to add one at this stage. Highlighting 'HISTOGRAM', for example, and then pressing <RETURN> will automatically add a **Histogram** subcommand to the **FREQUENCIES** command currently being edited and will also display the range of options available for this subcommand. One of these options is 'NORMAL'. If you highlight this and press <RETURN> then a **Histogram** subcommand will be added, specifying that a line showing a normal distribution curve for a population with the same mean value and standard deviation as the sample analyzed should be added to the output display.

Press **<Esc>** once more and you climb to the next layer of the menu system. This lists the various commands relating to 'Descriptive Statistics'. **FREQUENCIES** is included as one option (selecting this would take us 'back down' one layer), and others include **CROSSTABS** and **MEANS**.

Each time you press **<Esc>** you are either returned to the Main Menu or raised one step closer to the Main Menu. Thus pressing **Esc** in response to the 'Descriptive Statistics' menu raises us to the 'Analyze Data' layer, and a final press of **Esc** brings us to the main menu.

Thus we started our excursion in level 5 of the menu system and we have gradually climbed to the main menu. The steps followed the sequence 1 to 5 in the table:

5 Main Menu
4 Analyze data
3 Descriptive Statistics
2 FREQUENCIES
1 FREQUENCIES Statistics

This traces a single path in a menu system that has hundreds of different pathways. The menu system is time-saving and worth getting to know if you spend extended time analyzing data with SPSS/PC+ but for beginners and less frequent users, working with the menu system can be difficult and very confusing.

You can jump to the Main Menu at any time with **<Alt-Esc>**. However, on many occasions you will probably just wish to switch the menu system off and return to manual editing, and this can be achieved by pressing **<Alt-M>**.

10.8 Using the on-line glossary

SPSS/PC+ provides an on-line glossary which can be consulted only from within a **REVIEW** file. To look up the meaning of a term, enter the command **<F1><g>**. A prompt will appear: "Item to look up". If the editing cursor was on a word in the file you are editing then this word will be offered at the prompt as a suggestion (if the cursor was on the *first* word of a term such as 'standard deviation' in the file, then the whole term would be offered as a suggestion). If the offered term is the one you wish to have explained, then simply press <RETURN>. Otherwise, enter a term in reply to the prompt and then press <RETURN>. The term need not be a single word. 'Analysis of variance', for example, is acceptable as a single item to be looked up. If the term supplied is in the glossary (the glossary includes over 1200 terms) the corresponding definition will be displayed in a window at the top of the screen. If the word is not in the glossary then the closest term, alphabetically, is displayed, along with its definition. A line of text at the bottom of the window also provides information on the items to be found in the glossary immediately above and immediately below the current term. Definitions for these terms can be displayed by pressing **<Pg Up>** or **<Pg Dn>**, as appropriate.

Although the cursor remains within the editing screen, only limited movement is possible while a definition is displayed in the window. To switch off the window display, press **<Esc>**.

If you wish to keep a record of a definition displayed in the window, press **<F8>** and it will 'pasted' into the file currently being edited. The definition is written inside a box, making it separate from the rest of the file and easy to locate. Having returned to editing (by pressing **<Esc>** to switch off the window display), you may now wish to look up the meaning of a word which occurs within the definition. To do this, place the cursor on the word within the definition that you now wish to look up, and press **<F1><g>** again. If you wish to retain a record of this second definition you must enter the pasting command (**<F8>**) again, before entering **<Esc>** to return to editing.

Definitions can be useful in annotating the printed output. One way to achieve this is to edit the listing file during the session. At the SPSS/PC: prompt, enter:

```
SPSS/PC:review listing.
```

The cursor will initially be at the end of the listing file. You might wish to move the cursor to the top of the listing file (with **<Ctrl-Home>**) and then to search for the first occurrence of the problematic term using the **REVIEW** string search facility. Thus to find the first occurrence of the term 'mode' in your file, press **<F5><f>** and enter 'mode' (without apostrophes) at the 'Search string' prompt. Press <RETURN> and the first occurrence of 'mode' will be found. Now use **<F1><g>** to consult the on-line glossary. When the definition appears in a window at the top of the screen, press **<F8>** and the definition will be pasted into the listing file.

Save the edited file (**<F9><w>**), retaining the suggested name, and then exit from **REVIEW** (**<Alt-F10>**). When the session has finished, exit from SPSS/PC+, and rename the listing file. When you later edit the renamed file, using a word-processor, for example, you will be able to make use of any definitions which have been pasted in at relevant points within the file.

Appendix K provides a list of all of the **REVIEW** commands described in this book. There are also several others which you may wish to explore when you are more experienced in the use of SPSS/PC+. However, the commands described here are sufficient to enable you to edit files in as complete a way (though not necessarily in as efficient a way) as the system can provide, so you need not feel that your work with SPSS/PC+ will be limited in any way by your 'limited' knowledge of **REVIEW**.

11 *Statistical Analysis III: CORRELATION and PLOT*

In this chapter we consider the use of two further procedure commands, **CORRELATION** and **PLOT**.

11.1 The CORRELATION procedure

The **CORRELATION** procedure calculates the correlations between variables using the **Pearson product-moment** correlation formula. The procedure will tell us the degree to which pairs of variables are correlated.

A **positive** correlation means that high values of one variable (A) tend to be associated with high values of the other variable (B) and that low values of A tend to be associated with low values of B. We would expect, for example, that for the subjects in the Gender Attitudes Survey, the *problem* variable would be positively correlated with the *neur* (neuroticism) variable, and that *ht* (height) would be positively correlated with *wt* (weight). Incidentally, it would make no difference to the value of a correlation between measures of height and weight whether height was recorded in inches or centimetres, or whether the weight was entered as grams, pounds or kilograms.

A **negative** correlation means that high values of A tend to be associated with low values of B (and that low values of A tend to be associated with high values of B). We might expect, for example, that *var001* (the degree of agreement with the statement : **'Women are equal to men in every way'**) would be negatively correlated with *var004* (the degree of agreement with the statement : **'There are many activities in which men are clearly superior to women'**).

The value of a correlation is represented by the **correlation coefficient**. Negative correlations range from -1 to 0 (e.g. -.86 is a strong negative association, -.16 is a weak negative association). Positive correlations range from 0 to +1 (e.g. .78 represents a strong positive association, and .04 represents a very weak positive association). The SPSS/PC+ **CORRELATION** procedure displays correlation coefficients to four decimal places (e.g. -0.6354, 0.4653).

CORRELATION also displays the **significance** of each correlation coefficient calculated. It assumes that the user has hypothesized either a positive or a negative association and therefore normally displays the **one-tailed** probability of the correlation. If a directional hypothesis has not been made, the significance of the **two-tailed** probability can be requested by **Option 3**.

A single asterisk (*) indicates a one-tailed probability of less than 0.01. This means that a correlation of this level (or greater) would occur by chance (e.g. with repeated analyses of different subsets of random data) less than once in every 100 analyses.

A double asterisk (**) indicates a one-tailed probability of less than 0.001. This means that a correlation of this level (or greater) would occur by chance less than once in every 1000 analyses.

The Pearson product moment correlation makes certain assumptions about the kind of data used. A statistics textbook will provide details of these assumptions. The product moment correlation is used where the data (for *both* variables) are true numbers. To correlate codes or ranks, other forms of analysis are used (see Chapter 13).

Using the CORRELATION procedure

The command for a **CORRELATION** procedure takes the form :

```
SPSS/PC: correlation var001 with var002.
```

The normal listing rules and conventions hold, and multiple lists can be included in a single **CORRELATION** command. Thus :

```
SPSS/PC: correlation var003 to var006 with var007 var008
       : /var001 to var005 with var009 var010.
```

This requests 18 correlations altogether (8 are requested in the first subcommand and 10 in the second). The correlation values are displayed in the form of two rectangular matrices (one for each subcommand), with variables *before* the keyword **WITH** forming the *rows* (and those *after* forming the *columns*).

If the keyword **WITH** is omitted then correlations are performed between all pairs of the variables specified. Thus :

```
SPSS/PC: CORRELATION var003 to var006.
```

would produce a square matrix of (4 x 4) 16 correlation coefficients. However, six of these would be replications (e.g. *var003* would be correlated with *var004* AND *var004* would be correlated with *var003*, producing the same result on both occasions). Also, each of the four variables would be correlated with itself, producing a correlation coefficient of exactly 1 (displayed by **CORRELATION** as 1.0000). Of the 16 coefficients in the matrix, therefore, only six would be both unique and meaningful. If a correlation coefficient cannot be computed a period (.) is displayed at the appropriate point in the matrix.

The keyword **ALL** may be used, but this would be appropriate only if **all** of the variables were numerical (i.e. if the dataset included no codes).

The following command, specifying a number of variables created from the Gender Attitudes Survey data by the **COMPUTE** commands given on pp. 106–109:

```
SPSS/PC: correlation sexism extra neur problems.
```

produced the following display :

```
Correlations:  SEXISM       EXTRA        NEUR         PROBLEM

   SEXISM      1.0000        .0099        .0607       -.1913
   EXTRA        .0099       1.0000       -.2197       -.1151
   NEUR         .0607       -.2197       1.0000        .5641**
   PROBLEM     -.1913       -.1151        .5641**     1.0000

N of cases:     47         1-tailed Signif:  * - .01   ** - .001
```

Options for the CORRELATION procedure

Option 1 requests that missing values be *included.*

Option 2 specifies *pairwise deletion.* By default, any case with a missing value for *any* variable contained in a specified matrix is deleted from *all* the correlations in the matrix.

Thus, normally, if case 23 has a missing value for the variable *var006*, the following command :

```
SPSS/PC: correlation var001 to var008.
```

would exclude data for case 23 from all of the 64 correlations in the matrix (whether or not the particular correlation involved *var006*). Thus any cases with missing data for any of the specified variables will be excluded from all of the correlations in the matrix. The above display indicates that the analyses are based on the 47 cases (of the original 50) with valid data for each of the 4 specified variables. But if *pairwise deletion* were requested (**Option 2**), a case would be excluded only from those calculations for which it had missing data.

Option 3 specifies that the significance of *two-tailed* probabilities (rather than one-tailed) should be indicated by asterisks (* for 1 per cent probability and ** for 0.01 per cent probability).

Option 4 requests that the output matrices of the correlations be written to a *results file* (as well as displayed on the screen). Such results matrices can later be used as input for other procedures. Details on the use of this option are given on pp. 150–151.

Option 5 requests that, rather than simply displaying the **significance** of the correlation coefficient (with one or two asterisks, where * indicates a probability of less than 0.01 and ** indicates a probability of less than 0.001), the *exact probability* be displayed (e.g. probability = .0037). This option also causes the *count* (i.e. the number of cases used in the calculation) to be displayed for *each* correlation coefficient, and this aspect is particularly useful when **Option 2** (pairwise deletion) has also been requested.

Statistics for the CORRELATION procedure

Statistic 1 requests a display of the mean, the standard deviation and the count for each variable.

Statistic 2 requests cross-product deviations and covariance (see a statistics textbook for information on these measures).

N.B. Only one **Options** subcommand and one **Statistics** subcommand can be associated with any one **CORRELATION** command (although more than one option or statistic can be specified on each). Thus :

```
SPSS/PC: Correlation var001 to var006
       : /options= 2 5
       : /statistics= 1 2.
```

Writing results (matrix) Files

The output from the **CORRELATION** procedure can be written as a **results matrix** on a file which can later be used as **input** to another procedure (such as **REGRESSION**). Using such pre-processed data instead of the original data can save considerable processing time.

Option 4 allows a results file to be written from the results of the **CORRELATION** procedure. Such results must form a square matrix (i.e. every variable correlated with every other variable in the variables list). Thus subcommands which include the keyword **WITH** are ignored when matrices are written.

The correlation matrices are written using a standard format. Each coefficient occupies 10 columns (with 7 digits to the right of an implied decimal point). Thus each 80-column 'record' (a **record** is a line of data) can include only 8 values (although the rows of a matrix can extend over several records. This format will be expected by those procedures which may later use the matrix file as input.

Following the matrix of correlation coefficients, the results file will specify the number of cases on which the correlation coefficients are based. With listwise deletion (the default) a single value for *n* is given. With pairwise deletion, however, a separate *n* is given for each correlation coefficient. This is provided in a separate *n*-matrix which immediately follows the correlation matrix. All values of *n* are included in a 10-column format (this time with all digits to the **left** of an implied decimal point).

The name of the file to which the matrix is written is either "SPSS.PRC" (the default) or a name which has previously been provided on a **SET** command (using the **Results** subcommand). When naming a matrix results file the extension ".mat" is commonly used. Thus if it is intended to write a results file, a suitable name should be devised earlier in the session (before the **CORRELATION** procedure is started) and a **SET** command issued. For example:

```
SPSS/PC: set results = "GENDER.MAT".
```

The following command file will provide a name for the results file before including a data definition file, calculating a series of correlation coefficients and writing the matrix of results to a file:

```
set results= "gender.mat".
include "gender.def".
correlation var001 to var010
/options= 4.
```

This would produce (on a stored file named "gender.mat") a matrix of correlation coefficients with a single *n* value.

11.2 The PLOT procedure

The **PLOT** procedure produces a graphical display of variable frequencies – the frequencies of two variables are plotted simultaneously in a two-dimensional space. Although the procedure is very flexible, it can become rather complex, and the present discussion will therefore be confined to an initial exploration.

Plots and scattergrams

Let us consider two variables *var001* and *var002*. Any one of the respondents in a study will have a value (either a valid value or a missing value) for both variables, and *both* of these values can be displayed simultaneously as a single point on a **bivariate** (two variable) graph.

Suppose a single case has been given the value 3 for *var001* and the value 4 for *var002*. These values can be represented as a single point (**X**) on the following graph.

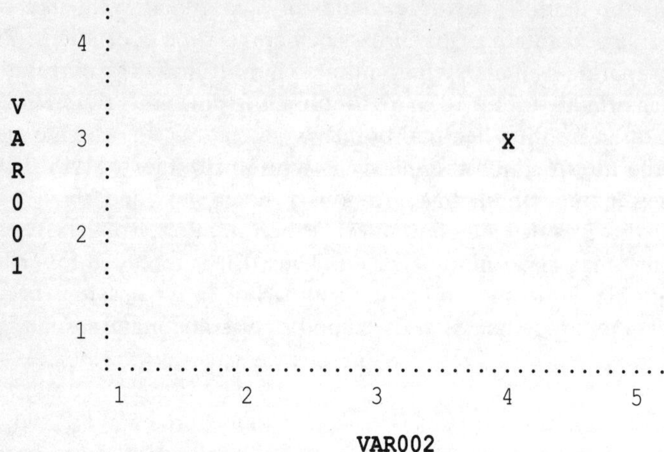

Now if we use the **PLOT** command to plot the data points for two variables (*var004* and *var007*) for the 50 subjects from the Gender Attitudes Survey, we obtain the following :

```
                  PLOT OF VAR004 WITH VAR007

        ++----+----+----+----+----+----+----+--+
   M     |              1         2    2    2  |
   e     |                                     |
   n    6|         1         1    1    1       +
         |                                     |
   s     |              2              4       |
   u     |                                     |
   p    4|    1    2    1    1    2    2       +
   e     |                                     |
   r     |         4    5    1    2    1    1  |
   i     |                                     |
   o    2|    1    3    1    2              1  +
   r     |                                     |
         |    1                        1       |
        ++----+----+----+----+----+----+----+--+
              1         3         5         7
         0         2         4         6

                  Women more sensitive
```

Here a 1 indicates a single data point, 2 indicates two data points, etc.

This type of graph is often referred to as a **scatterplot**, and we can see that the scatterplot is rather similar to a cross-tabulation table (in this case with 7 X 7 = 49 cells).

We can see from this scatterplot that there is some tendency for subjects to have either high values for both *var004* and *var007*, or low values for both of these variables. This suggests that there is a positive correlation between the two variables. This is confirmed by the correlation coefficient (0.4098, significant at the 0.01 level). Those who agree that men are superior in some activities also tend to agree that women are emotionally more sensitive than men.

Using the PLOT procedure

Within the procedure, the word **PLOT** has two separate functions. It acts both as the **command** word and also as an essential **subcommand**. Thus the word **PLOT** must occur twice (at least) for the procedure to work. The **PLOT** command may also include a number of other optional subcommands which affect the format of the graph and the labels and symbols used. Following any optional subcommands, the **PLOT** subcommand contains the list of the variables which are to be plotted in pairs.

The scatterplot of the Gender Attitude data, above, was produced using the simplest form of the **PLOT** command :

```
SPSS/PC: PLOT plot var004 with var007.
```

The above command contains none of the possible optional subcommands and therefore accepts all of the default characteristics. By default, a single two-dimensional scatterplot of the two variables is produced. The variable named first on the plot subcommand is displayed as the vertical axis. The output display includes the axes labelled with any variable labels previously specified. Cases with missing data are excluded from scatterplots in which data from one or both variables is missing, but are included in any other scatterplots requested on the plot subcommand.

The Optional Subcommands

Default parameters can be overridden by using a variety of optional subcommands (there is no **Options** subcommand for the **PLOT** procedure). Before examining some of the subcommands in detail, let us consider an illustrative example of how various subcommands appear within a single **PLOT** command.

In this case we are going produce a simple scatterplot of the computed variable *sexism* against one of its 'component' variables, *var004* (**'Men superior in many activities'**).

```
SPSS/PC: PLOT title= "sexism with 'Men Superior'"
       : /vertical= "sexism" reference (24)
```

```
: /horizontal= "MEN superior" reference (4)
: /plot sexism with var004.
```

We will first examine this command, statement by statement.

Title is the title given to the graph and is printed at the head.

Vertical in this case specifies the vertical **label** (i.e. the label, sexism, to be given to the vertical axis of the graph) and a **reference** point (here the *sexism* value '24'). This will cause a line will be printed on the graph at this point. Up to 10 such reference points can be specified for each of the vertical and horizontal axes.

Horizontal in this case specifies both a horizontal **label** (i.e. the label, MEN superior, given to the horizontal axis of the graph) and also a **reference** point, the value '4'.

Plot (subcommand) in this case requests a single 'bivariate' (i.e. two-variable) plot of *sexism* with *var004*. Lists of variables can be included in the plot subcommand in the usual way.

The above **PLOT** command produced the following output :

```
                    sexism with 'Men Superior'


         +|----|----|----|----+----|----|----|--+
         |                1    |                   |
         |                     |                   |
         |          1          |                   |
         |     1    1          |                   |
s     32+                1    1                   +
e        |          2    1    1                   |
x        |          1    3    |                   |
i        |                2    1                   |
s        |     1          3    |    1    1         |
m     24+--------------------+--------------------+
         |                1    2    4              |
         |          3    2    |         1    1    |
         |                     2              1    |
         |                     1    1         2    |
      16+                     |         1         +
         |                     |              3    |
         +|----|----|----|----+----|----|----|--+
               1         3         5         7
           0         2         4         6

                        MEN Superior
```

Notice that the scale of the axes has been adjusted automatically to conform to the page size.

The specified title is included as a heading (the default title in this case would have been **"Plot of sexism with var004"**).

Reference lines have been drawn; at the points requested, on the horizontal and vertical axes.

The output from this plot would seem to indicate that there is a strong association between responses to the 'men superior' item on the Gender Attitudes Questionnaire (coded as *var004*) and the composite *sexism* score. A **CORRELATION** analysis of these variables revealed a highly significant correlation (- .6558), indicating that a high *sexism* score is associated with a *low* score for *var004*. This indicates that 'sexist' respondents are more likely than 'non-sexist' respondents to agree with the statement that **'there are many activities in which men are clearly superior to women'**. However, it must be remembered that the **COMPUTE** command used to create the score for *sexism* included *var004* as one of the elements.

The MISSING subcommand

Normally, missing data are not included in the plot. Deletion is plotwise (equivalent to pairwise). But listwise deletion can be specified, using:

```
/Missing= listwise
```

and missing data can be included, using :

```
/Missing= include
```

The FORMAT subcommand

The **Format** subcommand specifies the type of plot produced. The default produces one or more **bivariate scatterplots**.

If **overlay** is specified then all variables specified on the next **Plot sub**command are plotted *together* on a single graph.

```
/Format= overlay
```

For an overlay plot, **PLOT** automatically selects a symbol for each 'layer' and also selects symbols to represent superimposed points. A key is provided on the display.

Other formats available include **contour** plots and **regression** plots (see *SPSS/PC+ 4.0 Base Manual*).

The SYMBOLS subcommand

The **Symbols** subcommand is used to specify which symbols should be displayed in the body of the graph. The default is alphanumeric. Thus 1 represents one data point, 2 represents two data points and so on. After 9 the symbols change to upper case letters A to Z. All values of 36 or above are represented by an asterisk (*). These symbols can be changed by the user. Thus, in the following subcommand :

```
/Symbols= "*"
```

the user is specifying that any number of data points (one or more) occurring at a particular point in the graph space should be marked with an asterisk. Here :

```
/Symbols= "*,@,?,&"
```

the user is specifying that an asterisk (*) should be used if there is one data point present, that @ should be printed if two data points are present, etc. Up to 36 separate symbols can be specified, and the printed plot will include a key.

The VSIZE and HSIZE subcommands

The **Vsize** and **Hsize** subcommands specify the vertical and horizontal sizes of the display frame used for the plot. Default values depend on the current page size of the paper used (either the SPSS/PC+ default or a size which has been set by a **SET** command). For normal 80-column printed pages the default values are **Vsize** = 16 (rows); **Hsize** = 38 (columns). Changes can be made in the following way:

```
/Vsize= 20  /Hsize= 45
```

The TITLE subcommand

The **Title** subcommand simply provides a title for a graph. Thus :

```
Title= "Sexism with 'Men Superior'".
```

The title can be up to 60 characters long.

The VERTICAL & HORIZONTAL subcommands

These subcommands can include a title for the axis. Thus :

```
/Vertical= "Sexism"
```

and they may also specify the minimum and maximum values to be plotted (default plots *all* values); any data which fall below a specified minimum or above a specified maximum are simply not plotted. Thus:

```
/Horizontal= "Men superior" MIN (2) MAX (6)
```

These subcommands can also be used to specify values at which **reference lines** will be printed, e.g. :

```
/Horizontal= "Sexism"  reference (24)
```

For overlay plots, when two or more variables with different ranges are to be plotted on the same axis, it is often useful to standardize scores so that each variable is effectively reduced to the same scale. This is done on the same subcommand, by specifying **standardize.** Thus :

```
/Horizontal= standardize
```

Finally, it is possible to specify that all plots on a **PLOT** subcommand list be plotted on axes with identical scales (normally the scales would differ, with each tailored to the particular data being plotted). This is achieved by specifying **uniform**. Thus:

```
/Horizontal= uniform
```

It can be seen that rather a lot of information could be contained on a single **/Horizontal** or **/Vertical** subcommand. Thus :

```
/Vertical= "Salary"  MIN (6000) MAX (10000) standardize
```

N.B. The subcommands **Missing**, **Vsize**, **Hsize** and **Symbols** apply to **all** plots within a **PLOT command** and can therefore be specified only **once** in any **PLOT** command.

If you find all these subcommands (and there are others) intimidating, take heart. For most purposes the defaults will probably serve you well. Remember that we produced an interesting scatterplot with just :

```
SPSS/PC: PLOT Plot = var004 with var007.
```

Multiple plots can be specified by a single command. For example:

```
SPSS/PC: PLOT PLOT= var001 with var002;
       : var003 var004 to var 006 with var010;
       : var008 with var009.
```

(Notice that a semi-colon, not a slash, is used). The above command requests six bivariate plots.

Although, as in this example, several **Plot sub**commands can be included in a single **PLOT** command, some subcommands (such as **Title**, **Horizontal**, **Vertical** and **Format**) affect only the **Plot** subcommand which immediately follows them. Therefore, even if default characteristics are to be changed in exactly the same way for successive **Plot** subcommands, the appropriate changes must be specified *each* time.

Thus, to request different titles, but the same reference lines, for two successive **Plot** subcommands, a command of the following kind would be needed:

```
SPSS/PC: PLOT  / title= "First two variables"
       : /horizontal= reference (4)
       : /vertical= reference (3)
       : /plot= var001 with var002
       : /title= "Second two variables"
       : /horizontal= reference (4)
       : /vertical= reference (3)
       : /plot = var003 with var004.
```

This one **PLOT** command statement contains two separate **Plot** subcommands (and six additional subcommands). The final subcommand must always be a **Plot** subcommand.

A special pairing can be requested (similar to the special pairing we met in **T-TEST Option 5**), by specifying **(pair)** at the end of a list containing the same number of variables on either side of the **WITH** keyword. Thus, for example:

```
SPSS/PC: PLOT plot age with var001 to var006 ;
       : var003 to var006 with var007 to var010 (PAIR);
       : var008 with var009 var010.
```

Here *var003* would be paired *only* with *var007*, *var004* would be paired only with *var008*, and so on. The complete command would therefore produce 12 plots.

No **Statistics** subcommand is available for the **PLOT** procedure.

12 Statistical Analysis IV: ONEWAY and ANOVA

Both of the procedures **ONEWAY** and **ANOVA** perform analysis of variance (anova). This statistical test compares the values of a dependent variable for cases which fall into a number of different groups. The independent (between subjects) **T-TEST** procedure does the same thing, but only two groups can be compared in each *t*-test analysis. A one-way analysis of variance can compare two, three or more groups classified in terms of a single independent variable. A two-way analysis of variance compares groups formed by classifying cases simultaneously on two independent variables. If the SPSS/PC+ analysis of variance is to include just one independent variable a one-way anova is performed using either **ONEWAY** or **ANOVA**. With two independent variables a two-way anova is performed; and so on for three-way anova, four-way anova, etc. For an anova with two or more (and up to 10) independent variables, the procedure **ANOVA** must be used.

As well as being a method for the statistical analysis of data, anova refers to an approach to experimental design. Many studies are devised to conform to a particular type of anova, so that there is a close association between the nature of the study and the analysis which is eventually performed on the data obtained. The procedures **ONEWAY** and **ANOVA** are appropriate for designs in which **different cases** are found within each of the conditions. This is therefore similar to the **independent *t*-test**. There are anova designs in which comparisons can be made between different variables for the **same set of subjects** (as in the **paired** *t*-test). Such 'repeated measures' designs cannot be analyzed by the procedures detailed in this Chapter, however. If you wish to analyze data from a study which follows such a design you should use the **MANOVA** procedure which is included within the *SPSS/PC+ Advanced Statistics* optional module.

ONEWAY and **ANOVA** perform an analysis of variance to assess how much of the variation (variance) of a dependent variable can be assigned to different sources, including the independent variable(s) and any interactions between such variables. For example, it might be found that a measure of psychological difficulties, *problem* (the dependent variable), varies systematically with two independent variables, *sex* and *faculty*. These variables would therefore be seen as significant factors in determining the scores of the dependent variable. It might also be found that there is a significant *sex* X *faculty* interaction such that, for example, high *problem* scores tend to be found for male science students and female arts students.

Anova calculates the variance of the scores *within* conditions (for example, within *each* of the three faculties) and *between* conditions (in this example, between the three different faculties). If the *within*-conditions variance is low compared to the *between*-conditions variance then it would seem that the scores for the dependent variable tend to be different in the different conditions. This would suggest that the independent variable is able to explain a significant part of the overall variation in the scores.

Anova partitions the **Total Variance** into the **Treatment Variance** (i.e. that which is explained by the independent variables and any interactions between them), and the **Residual Variance** (i.e. that which is *not* explained) and compares these two values. If a substantial part of the total variance relates to differences between the conditions of an independent variable then that variable would seem to explain a significant part of the variance. The effect of this variable would therefore be seen as significant (i.e. it would have a low chance probability).

Taking the simplest example, the one-way anova, the stages in the computation can be outlined in the following way. First the between-groups and within-groups sums of squared deviations from the mean are calculated. Each of these values is then divided by the relevant degrees of freedom to yield a mean square value, and when the between-groups mean square is divided by the within-groups mean square, a variance ratio (F-score) is obtained. If this value is sufficiently high (corresponding to a *low* probability) then it will be statistically significant, and the effect of the relevant independent variable will have been shown to be significant.

12.1 The ONEWAY procedure

A one-way anova deals with only one independent variable. The variance *between* the groups formed by this independent variable is compared with the residual variance – i.e. the variance *within* the groups.

The simplest form of the **ONEWAY** command is:

```
SPSS/PC: ONEWAY var003 BY fac(1,3).
```

Here *var003* is the dependent variable. The numerical values of *var003* will be compared for different groups of cases (subjects). There is no requirement that these groups be of equal size. The groups will be formed from the categories corresponding to particular values of *fac* between 1, which is specified in the command as the *minimum* value to be used in forming groups, and 3, which is specified as the *maximum* value to be used. The minimum and maximum values must be integers (i.e. whole numbers, without a decimal point).

Because of the nature of the statistical test to be performed, only one independent variable may be specified in a single **ONEWAY** subcommand (if you wish to perform a similar analysis with more than one independent variable then the alternative procedure **ANOVA** should be used).

Although only one independent variable can be specified on a **ONEWAY** subcommand, up to 100 dependent variables may be specified. The normal variable listing conventions are permitted. Thus:

```
SPSS/PC: ONEWAY var001 to var010 BY fac(1,3).
```

This would produce 10 one-way analyses of variance. Each of the 10 dependent variables would be compared across the three *fac* groups. An analysis of variance will tell us whether there is a significant variation in the dependent variable (say, *var001*) between the three *fac* groups analyzed.

To produce the following example, a new variable *sextract* was computed from the Gender Attitudes Survey data. The **COMPUTE** command was the same as that for *sexism* (p. 109) except that it did not include *var003* as an element. The raw scores for *sextract* were then recoded to form three groups, 'low *sextract*', 'medium *sextract*' and 'high *sextract*':

```
SPSS/PC: recode sextract (10 thru 18=1) (19 thru 23=2)
       : (24 thru 35=3).
```

The **ONEWAY** procedure was then used to see whether the three *sextract* groups differed significantly on the variable which had not been used in the computation of the original *sextract* score:

```
SPSS/PC: oneway var003 by sextract(1,3).
```

This produced the following output:

```
            - - - - - - - - - - - O N E W A Y - - - - - - - - - - -

       Variable  VAR003     Women use sexual dis. as excuse
    By Variable  SEXTRACT   Recode 10 to 18-19 to 23-24 to 35

                              Analysis of Variance

                             Sum of        Mean            F      F
         Source      D.F.   Squares      Squares        Ratio  Prob.

Between Groups         2     20.0761     10.0380       4.1124  .0229

Within Groups         45    109.8406      2.4409

Total                 47    129.9167
```

Because there are 48 valid cases (*N*) the value of the 'total degrees of freedom' is (*N*–1) 47. Because there are three groups (*k*) the 'between-groups degrees of freedom' is 2 (i.e. *k*–1). The difference between these values (47–2 = 45) gives the degrees of freedom assigned to the residual ('within-groups') variance. The sums of squares have been calculated for the whole population (total sum of squares) and for the between-groups and within-groups sources. Dividing the between-groups and within-groups sums of squares by their respective degrees of freedom has yielded a mean square value for each. Dividing the between-groups mean square by the within-groups mean square has produced an *F*-ratio of 4.1124, corresponding to a probability of 0.0229. There is thus a significant effect.

With no **Options** or **Statistics** subcommands in effect, the default table is produced, as above, for each anova requested. The table contains the *F*-value and its probability. It also includes the degrees of freedom and both the sum of squares and mean squares for each of the two sources (within-groups and between-groups). Although the significance of the overall variation between groups is given by the *F* statistic, no tests to determine differences between individual pairs of groups have been performed. Cases with missing values on either the independent or dependent variable are excluded (pairwise).

A number of subcommands may be included within the **ONEWAY** command. **Ranges**, **Statistics** and **Options** will be described. Two of the subcommands available for **ONEWAY** (**Polynomial** and **Contrasts**) will not be discussed here.

The Ranges subcommand

When an independent variable (such as *sexism*, recoded into three categories) is shown to have a significant effect, we know that there is an overall difference in the values of the dependent variable across the groups. We do *not* know, however, which *pairs* of groups differ significantly. This can be assessed using a range test which can be requested using the **Ranges** subcommand.

Ranges specifies one of seven available tests to be used in identifying which pairs of groups (within the range specified for the independent variable) have significantly different means.

If three groups are specified for the independent variable then three comparisons will be made. Comparisons will be made between:

```
Group 1 and Group 2; Group 1 and Group 3; Group 2 and Group 3
```

```
• 4 groups would produce  6 comparisons
• 5 groups would produce 10 comparisons
• 6 groups would produce 15 comparisons
• etc.
```

The range tests available include the Duncan multiple range test (Duncan), the Least significant difference (LSD) and the Student-Newman-Keuls (SNK).

The default alpha (significance level) for all range tests is 0.05, but for certain tests (including Duncan and LSD) this level may be changed.

The **Ranges** subcommand is used in the following way:

```
: /ranges=duncan
: /ranges=lsd(.01).
```

Each range test has to be specified, as above, on a separate **Ranges** subcommand. For the first range test specified (Duncan), the default alpha level (0.05) is accepted. For the second test (LSD), however, the value is changed to 0.01.

In the range test output, the means of the groups are listed in ascending order and a grid is produced, with asterisks used to show which means are significantly different. Thus, in the analysis of *var003* by *sextract*, the following command:

```
SPSS/PC: oneway var003 by sextract(1,3)
       : /ranges=duncan.
```

produced a **Ranges** output which included the following Table:

```
(*) Denotes pairs of groups significantly different at the .050
     level

     Variable  VAR003     Women use sexual dis. as excuse

                          G G G
                          r r r
                          p p p

    Mean       Group      3 1 2

    3.6000     Grp 3
    4.8421     Grp 1      *
    5.1429     Grp 2      *
```

This output indicates that the mean values of *var003* for Groups 1 and 3, and 2 and 3, are significantly different but that there is no significant difference between the mean values of this variable for Groups 1 and 2 on the range test used.

Thus the 'high *sextract*' group differs significantly from both the 'medium *sextract*' groups and the 'low *sextract*' group in the degree of agreement with the statement **'Many women use sex discrimination as an excuse for their own shortcomings'**. From the Group means displayed in the table we can see that the 'high *sextract*' group (Group 3) has a *lower* mean. Because of the way in which the data were coded, this means that those who are highly sexist, as judged by their responses to other statements, also tend to *agree* more with the above statement.

OPTIONS

A number of options are available for the **ONEWAY** procedure. Some affect the way in which missing values are treated. Thus **Option 1** specifies that missing values should be included in the analysis, and **Option 2** specifies that missing values should be excluded listwise. **Option 3** requests that variable labels should be suppressed, and **Option 6** specifies that value labels should be used to label the groups. Three of the **Options** deal with the reading and writing of matrices.

Matrices

In Chapter 11 it was shown how output from the **CORRELATION** procedure could be written as a results file (by specifying the **CORRELATION Option 4**). The **ONEWAY** procedure can also produce a results file, and it can make use of results files produced in previous analyses.

Writing results files

To request that an output matrix from the **ONEWAY** procedure be written to a separate results file, **Option 4** is used. The default name for such a file is "SPSS.PRC", but a **SET Results** command can be used to specify an alternative name (the extension **.mat** is often used). The following command sequence (using *sexism*, recoded in the following way – 14 thru 20 = 1; 21 thru 26 = 2; 27 thru 38 = 3) was used to execute a **ONEWAY** procedure and then to write a results file:

```
SPSS/PC: SET Results = "GENDER.MAT".
SPSS/PC: oneway var007 BY sexism(1,3)
       : /OPTIONS= 4.
```

The **SET Results** command arranges for the results of the procedure which follows to be written to a results file named "gender.mat". The **ONEWAY** command requests an analysis of variance of the values of the dependent variable (*var007*) for three groups defined by the specified values of the independent variable *sexism*. The **Options** subcommand requests that the results matrix be written to the file named on the **SET Results** command.

The above command produced the following "gender.mat" file:

```
14.00      17.00      17.00
 5.0000     4.0000     3.5294
 1.5689     1.8371     1.8068
```

This results matrix contains 'vectors', for each group, of:

```
Line 1     The count
Line 2     The mean
Line 3     The standard deviation
```

All vectors are written over 10 columns, with two places to the right of the decimal point for the counts, and four places to the right of the decimal point for the means and standard deviations.

Because 10 columns are occupied by each vector, the maximum number of vector values (one per group) which can be recorded on a single line (a record) is eight. If more than eight groups are involved in the analysis then the matrix will include two or more records for each vector.

Reading matrices

When a results matrix has been stored as a results file it can be used to provide the input for certain procedures. A suitable results file can be read into **ONEWAY** by specifying the name of the file and including the relevant **Options** subcommand (in this case, **Option 7**):

```
SPSS/PC: data list file= "gender.mat" matrix/ var007 sexism.
SPSS/PC: oneway  var007 by sexism(1,3)
       : /options= 7.
```

In this example, the **DATA LIST** command specifies the name of the file from which the input is to be taken. It also informs the system that this is a matrix file and supplies names for both the dependent variable and the independent variable to which the matrix information relates. The **Options** command informs the system of the type of information included in the matrix.

If these commands followed the **ONEWAY** procedure described above, they would cause the system to read the results file and repeat the original analysis, but this time using the matrix information rather than the original data. New names could be given to the variables read from the file. The results produced would, of course, be the same as those previously obtained.

In the **ONEWAY** procedure, two **Options** are available for reading matrices, and these lead the computer to 'expect' the matrix to contain information in one of the two following arrangements:

Option 7 leads the computer to expect a matrix in the same format as that which is written by specifying **ONEWAY Option 4** (i.e. counts, means and standard deviations in the 10-column format specified above).

Option 8, however, leads the computer to expect *four* vectors: *counts* (for each group), *means* (for each group), *pooled variance* (within-groups mean square: one value only), and *pooled variance degrees of freedom* (one value only).

STATISTICS

Additional statistics can be obtained by using the **Statistics** subcommand. Three sets of additional statistics are available.

Statistic 1 requests further descriptive statistics for each group, including the minimum, maximum, mean, standard deviation and standard error. The other available **Statistics** request that complex parameters of the data be tested.

A **ONEWAY** command should have no more than one **Options** and one **Statistics** subcommand (although a number of options can be requested on each). Only one independent variable can be specified, and no more than 100 dependent variables.

12.2 The ANOVA procedure

The **ANOVA** procedure is used to obtain analyses of variance for a numerical dependent variable on up to 10 categorical (classification) variables (called 'factors'). Thus up to a '10-way' analysis of variance can be performed. There must be different subjects (cases) in each of the conditions. Analyses of variance can take several forms (factorial, hierarchical, regression), each of which can be specified by a suitable **Options** subcommand.

In this introduction only the simpler uses of the procedure will be discussed. Those who are unfamiliar with analysis of variance will need to consult a statistical text.

The simplest form of the **ANOVA** command is:

```
SPSS/PC: ANOVA var007 BY sexism(1,3).
```

This would produce a one-way analysis of variance, with the re-coded *sexism* as the independent variable and *var007* as the dependent variable. Three independent groups are specified (those taking the *sexism* values 1, 2 and 3).

In the following **ANOVA** examples, the personality variables *extra* and *neur* were recoded from the original values (-6 to +6) into three values, using the following **RECODE** command:

```
SPSS/PC: recode neur extra (lo thru 0 = 1) (1 thru 3 = 2)
       : (4 thru hi=3).
```

The command:

```
SPSS/PC: ANOVA var005 BY sex(1,2) neur(1,3).
```

requests a two-way anova with *var005* as the dependent variable. This command produced the output reproduced below.

```
          * * * A N A L Y S I S   O F   V A R I A N C E * * *

              VAR005   Women's movement ineffective
          BY  SEX      Subject sex
              NEUR     Neuroticism - coded 3 groups.

                              Sum of              Mean                Signif
Source of Variation          Squares      DF     Square         F      of F

Main Effects                  34.096       3     11.365     5.806      .002
  SEX                         15.488       1     15.488     7.912      .007
  NEUR                        13.518       2      6.759     3.453      .041

2-way Interactions             1.693       2       .846      .432      .652
  SEX       NEUR               1.693       2       .846      .432      .652

Explained                     35.789       5      7.158     3.657      .008

Residual                      84.170      43      1.957

Total                        119.959      48      2.499
     50 Cases were processed.
      1 CASES (  2.0 PCT) were missing.
```

This table indicates that both of the independent variables *sex* and *neur* have a significant effect on the dependent variable *var005*. The two sexes and the three *neur* groups differ in the degree to which they agree with the statement '**The women's movement has had little serious effect**'. However, there is no significant interaction between *extra* and *neur* with regard to the level of agreement with the statement. From an output of the means for each cell (requested by **Statistics 3**) it was found that the subjects who disagreed most with this statement were females, and those who obtained a relatively low score for neuroticism.

Up to five *dependent* variables can be specified on any one subcommand (and five subcommands can be included in a single command – subcommands should be separated by slashes /) e.g.:

```
SPSS/PC: ANOVA var007 to var010 BY neur(1,3)
       : extra(1,3) sex(1,2)
       : /var006 BY neur(1,3) sex(1,2).
```

would produce four three-way anova tables for the first subcommand and a single two-way anova table for the second.

By default, for less than six factors, a full factorial anova is used, with the main effects assessed before the interactions. All interactions are displayed.

The anova table includes a count of cases (and a separate count of missing cases), sums of squares, degrees of freedom, the mean square, F-value and the probability of F for each of the effects. Variable labels are also included. Missing cases are deleted listwise.

OPTIONS

Altogether, 11 **Options** are available for the **ANOVA** procedure. The following affect the way in which missing values are handled, the display of labels, the print format and the suppression of interactions.

Option 1 specifies that missing values should be included in the analysis.

Option 2 requests that labels should not be displayed.

Option 3 requests that interactions should be suppressed.

Options 4, **5**, and **6** request the suppression of 3-way, 4-way and 5-way interactions (respectively).

Options 7 and **8** deal with actions to be taken with regard to **covariates**

Options 9 and **10** specify alternative methods (Regression – **Option 9**, or Hierarchical – **Option 10**) to be used in the analysis of variance. See the *SPSS/PC+ Statistics 4.0* manual for details.

Option 11 changes the display and print format to narrow.

STATISTICS

Three additional **ANOVA** statistics are available.

For details on **Statistics 1** and **Statistics 2** consult the *SPSS/PC+ Statistics 4.0* manual.

Statistic 3 requests a table of the counts and means for each cell. Thus with two independent variables, one (such as *fac*) specifying three groups, and the other (such as *sex*) specifying two groups, there will be six cells (male arts, female social studies, etc.). The values and counts for the rows and columns of this table are also displayed.

We have now completed our discussion of the procedures **ONEWAY** and **ANOVA**. The use of **ANOVA**, in particular, demands a sophisticated appreciation of statistics, and it should be remembered that anova is a method of experimental analysis rather than simply a statistical test. In a real study, anova should not be used in a *post hoc* fashion, as in the examples given, but should be built into the design of the experiment. As with other methods of analysis, your use of the SPSS/PC+ **ANOVA** procedure should not extend beyond your level of statistical understanding.

13 *Statistical Analysis V: Non-parametric Statistics – NPAR TESTS and RANK*

The **NPAR** procedure allows the user to carry out a variety of non-parametric statistical tests. Unlike parametric statistics (including the *t*-test, anova, the product-moment correlation, etc.) the use of non-parametric tests involves few assumptions about the nature of the data. They may be useful when there are relatively few cases in the sample we wish to examine, or where a frequency distribution is highly skewed (rather than symmetrical). Whereas a procedure like **ANOVA** demands that the dependent variable be truly numerical, non-parametric tests can handle data at other 'levels of measurement'. In particular they allow us to analyze 'nominal' data and 'ordinal' data.

Nominal data are merely classificatory. For example, players in a sports team will play in different positions. They may even have numbers on their shirt indicating their position. In a room where members from many teams were assembled we could count the Number 1's, the Number 2's, etc. We might like to know whether significantly more Number 1's than Number 2's were attending the meeting. But it would be quite inappropriate to think of the numbers indicating the players' positions as real numbers. It would be outrageous, for example, to suggest that 5 Number 1's are in some way equal to one Number 5. The numbers sewn to the players' shirts act only as labels (symbols or letters of the alphabet could be used instead), and cannot be used meaningfully as elements in an addition, subtraction, multiplication or division. They are merely nominal. Social class and sex are examples of categorical variables. When *class* or *sex* are used, in **ANOVA**, for example, they always appear as independent variables, never as dependent variables. For some non-parametric tests, however, social class (coded, perhaps, 1 to 5) and sex (coded 1 and 2) could be used as **dependent** variables. *Sex* is an example of a **dichotomous variable**, since all cases must fall into one of only two possible classes (data obtained from coin-tossing, which must all be either 'heads' or 'tails', provide another example).

Ordinal data are rank-ordered. The winner of a marathon is 1st, the runner-up is 2nd, and others who finish the race come 3rd, 4th, 5th, etc. The runners may start out with numbers sewn to their shirts, but these are merely labels for identification (they are 'nominal'). The positions in which they finish, however – 1st, 2nd, etc. – are not just labels. The numbers mean something. We know that the person who came 31st, for example, must have run a faster race than the person who came 38th. But we *cannot* assume that there was an equal time-gap between the 1st and the 2nd, and the 2nd and the 3rd, etc. Neither can we add the numbers (making the 2nd plus the 3rd somehow equal the 5th).

Various non-parametric statistical tests allow us to draw meaningful inferences from data which are at the nominal or ordinal level of measurement. The SPSS/PC+ procedure **NPAR** does not specify a particular test. Fifteen different non-parametric tests are available and the appropriate test must be specifically requested.

13.1 Choosing the appropriate test

The choice of test depends on a number of factors:

- the nature (level of measurement) of the data (e.g. 'nominal' or 'ordinal')
- the number of samples or groups (1, 2 or k)
- whether the two (or more) samples are related or independent
- what we want to do with the data (e.g. assess the level of association between two variables, determine differences in the central tendency or span of two or more groups, examine sequence patterns, or examine proportions within a single group)
- the preference of the user (i.e. sometimes a number of alternative tests will be suitable for assessing some aspect of the data).

Table 13.1 presents a concise account of the non-parametric tests available within the **NPAR** procedure. The tests are listed in the order in which they are discussed in the rest of this chapter. The sequence is organized in terms of the number of samples, whether the samples are independent or related, and whether nominal or only ordinal level data can be analyzed by the test.

The levels of measurement (nominal, ordinal and, for parametric statistics, interval) can be thought of as occupying different levels on an ascending scale from nominal to interval. A nominal data test (such as the chi-square test) can be used to analyze data at *any* level. An ordinal test can be used only for data which is *at least* at the ordinal level. The general rule is that a test which is designed to analyze one level of data can be used to analyze data at that level or at any *higher* level, but cannot analyze data which is at a *lower* level.

Table 13.1 Non-parametric tests available within the NPAR procedure

Test	NPAR Name	N nominal O ordinal D dichot.	Purpose	No. of samples	Ind. or Related
1 Chi-square	CHISQUARE	N	Tests fit between observed and expected frequencies	1	–
2 Binomial	BINOMIAL	N (D)	Tests for biased proportion	1	–
3 Runs	RUNS	N (D)	Tests whether run sequence is random	1	–
4 Kolmogorov-Smirnov	K-S	O	Tests match between frequencies	1/2	I
5 Wald-Wolfowitz	W-W	O	Tests for diffs in ranks of ind. groups	2	I
6 Moses	MOSES	O	Compares ranges of two Indep. groups	2	I
7 Mann-Whitney	M-W	O	Tests for differences between Indep. groups	2	I
8 McNemar	MCNEMAR	N (D)	Tests before/after change in D variables	2	R
9 Sign	SIGN	O	Tests paired data for + or – bias	2	R
10 Wilcoxon	WILCOXON	O	Tests for differences between Related groups	2	R
11 Median	MEDIAN	O	2-way frequency association	2–k	I
12 Kendall	KENDALL	O	Agreement between judges	2–k	R
13 Kruskal-Wallis	K-W	O	1-way anova	k	I
14 Cochran	COCHRAN	N (D)	Tests before/after change in k groups	k	R
15 Friedman	FRIEDMAN	O	2-way anova	k	R

13.2 One-sample tests

The chi-square test – NPAR Chisquare

The chi-square test available within the **CROSSTABS** procedure examines the frequencies of nominal data as they are crosstabulated for the values of two or more variables. **NPAR Chisquare**, however, employs the chi-square test in a different manner, to see whether the observed frequencies of data across the values of a single variable conform to an 'expected' pattern.

The output display includes the observed and expected frequencies (with the differences, or 'residuals'), and the chi-square value, with the related degrees of freedom and significance.

All or part of the value-range for the variable can be tested for the 'fit' between the observed and the expected frequencies. By default, equal frequencies are expected for all of the values.

Suppose that we asked 1000 children to choose their favourite flavour of ice-cream from a range of 9 flavours. We could code their choice for the variable *prefice* using a 1 to 9 coding system. If we wished to test the assumption that all of the flavours would enjoy equal popularity, then we could rely on the default ('equal') and would not need to specify particular expected frequencies. The command used to request a one-sample chi-square test in this case would be:

```
SPSS/PC: NPAR Chisquare = prefice.
```

If, however, we had some reason to believe that the strawberry flavoured ice-cream might be twice as popular as any other flavour, we could test the observed preferences against the expected frequencies. In the following example we use a chi-square test to find out whether the observed frequencies match or 'fit' the expected frequencies of the five flavours (coded 1 to 5). Thus the example also shows how we can examine the frequency match for only *some* of the values originally coded.

We expect two children to have chosen strawberry flavoured ice-cream (coded 5) for every one child who has chosen *each* of the four other flavours. This could be tested with the command:

```
SPSS/PC: NPAR Chisquare=prefice(1,5) / EXPECTED = 4*1, 2.
```

We see from this example that the expected frequencies are interpreted as **proportions** rather than as absolute frequencies. Notice also the convention (in this example '4*1') used to specify a uniform expected frequency for a range of values. The expected frequencies must be listed in order of ascending value (in this example, from value 1 to value 5).

NPAR Statistics and **Options** are discussed in Section 13.7.

The binomial test – NPAR Binomial

The binomial test determines whether a variable which takes one of only two possible values (i.e. a dichotomous variable) is randomly (binomially) distributed. By default the test assumes that either of the two possible values is equally likely (i.e. a 0.5 probability is assigned to each value) although this can be changed. The output display includes the count of valid cases, the proportion of cases taking the first value and the two-tailed probability of this proportion. If the probability is low, and reaches significance level, then we can say that the two values of the variable are *not* randomly (binomially) distributed.

A binomial test could be used to determine whether a coin (which can be tossed repeatedly to take a heads or tails value for each toss) is biased – i.e. whether the probability of the observed heads and tails frequencies differ significantly from 0.5. Using the stereotype variables from the Gender Attitudes Survey (*st01* to *st10*) we can determine whether the proportion of respondents who felt that more men or more women would agree with each of the 10 statements was significantly different from a 50:50 split.

A binomial test to determine whether *st05* was biased could take the form:

```
SPSS/PC: NPAR BINOMIAL = st05 (1,2).
```

Because each of the stereotype variables takes only two valid values, no values need be specified. A list of variables can be included. The following command illustrates both of these points:

```
SPSS/PC: NPAR Binomial = st01 to st10.
```

The output from this command included the following:

```
- - - - Binomial Test

   ST01      Depressed often (M-W)

    Cases
      48    =   2            Test Prop. =  .5000
       1    =   1            Obs. Prop. =  .9796
      --
      49   Total             Z Approximation
                             2-tailed P =  0.0
```

From this we can conclude that the values for this variable are not randomly distributed between the two possible values. We can see that the respondents believed that more WOMEN than MEN would agree with the statement **'I often feel a little depressed'**.

The results of the above command also revealed that there was a significant stereotyping effect on several of the other variables. Further results indicated that women were also believed to be more likely than men to report feeling anxious and having sleep difficulties. Men, on the other hand, were seen as more likely to report thinking about sex a lot, enjoying taking risks and solving abstract problems, and trying to dominate people.

The binomial test can also be used when the prior probability of a particular value is not 0.5. The following example tests the hypothesis that a die is biased such that the value 1 is presented with a frequency significantly greater than 1/6 (0.167). The results of successive throws are coded 1 to 6 for the variable *throw* and the frequency of 1s is tested against the perfectly unbiased expected frequency:

```
SPSS/PC: NPAR binomial(.167) = throw(1).
```

The 1 in brackets after the variable name specifies that the value 1 is the first value of the dichotomy and that *all other values* (in this case 2 to 6) are treated together as the second value (which would have the expected frequency of 0.833, or 83.3%).

If the default probability (.5) is changed then the output gives the *one-tailed* probability.

The runs test – NPAR Runs

The runs test is used to determine whether, for a dichotomized variable (such as heads and tails, or male and female) the *sequence* of values across cases is random. Thus we could test whether, in a two-party state, there is a tendency for one party to remain in office through several successive elections, until the other party is elected (after which it, too, stays for several terms). We could test this by analyzing historical data, where each case is a term in power (the variable *govmnt*) and the two parties are coded 1 and 2. Clearly, the order in which the cases were placed would be of the utmost importance (in this example it would have to be the chronological order of the successive governments). The appropriate **NPAR Runs** analysis could be requested using the following SPSS/PC+ command:

```
SPSS/PC: NPAR Runs(2) = GOVMNT(2).
```

The (2) specifies the cutting point. Values *below* this are treated as being on one side of the dichotomy, and values equal to or greater than this point are treated as being on the other side.

The display includes the cutting point, the count for each side of the dichotomy, the *Z* statistic with its one-tailed probability, and the number of runs – a run is an unbroken sequence of one value. Thus, in these examples, list (a) contains 8 runs and list (b) contains 21 runs:

(a) 000011011111101111111111101111111111
(b) 00110100011101010011101110110101110

Variables can be dichotomized either by specifying a numerical cutting point, as in the command example above, or by ordering that the cutting point be at the mean, the mode or the median, e.g.:

```
SPSS/PC: NPAR Runs(median) = test1.
```

In the following example, nine variables are subjected to the **Runs** test in 3 subcommands:

```
SPSS/PC: NPAR Runs(mode) = item1 item9/ Runs(mean)= item2
       : to item6/ Runs(18.99)= score4 score7.
```

13.3 One- or two-sample tests

The Kolmogorov-Smirnov tests – NPAR K-S

The Kolmogorov-Smirnov (K-S) test is used either to compare the frequency distribution of one sample against a theoretical distribution (the **one-sample test**) or to compare the frequency distributions of two independent samples (the **two-sample test**).

One-sample test
The one-sample **K-S** test is a test of the 'goodness of fit' between a distribution of observed frequencies and a theoretical frequency distribution. The theoretical distributions available as templates for matching are the *normal* distribution, the *Poisson* distribution and a *uniform* (or rectangular) distribution – i.e. equal frequencies for all values. The value computed is Z, a measure of how much the observed distribution differs from the theoretical.

By default, the display includes the count of valid cases, the range, the Z value (with its two-tailed probability) and the most extreme differences between the observed and theoretical frequencies.

The theoretical distribution with which the observed frequencies are to be compared must be specified in brackets after the K-S characters (e.g. **(normal)**).

- for the **normal** distribution a mean value of the theoretical distribution and its standard deviation can be specified
- for the **Poisson** distribution a mean can be specified
- for the **uniform** distribution, minimum and maximum values can be specified.

Such additional parameters are optional (by default these values are taken from the observed frequency distribution).

The following example tests whether the observed frequencies of grades (from A to L) given to students (the variable *grade*, coded 01 to 12) conform to a **uniform** distribution (i.e. whether approximately equal numbers of students receive each of the 12 grades):

```
SPSS/PC: NPAR K-S(UNIFORM) = grade.
```

The next example tests whether the frequency distribution of IQ scores obtained from a population of children in a large national survey conforms to the theoretical distribution – a **normal** distribution with a mean of 100 and a standard deviation of 15:

```
SPSS/PC: NPAR K-S(NORMAL, 100,15) = IQ.
```

A non-significant probability value in the output display indicates that that distribution does *not* depart significantly from a normal distribution with the specified mean and standard deviation.

You could use the command:

```
SPSS/PC: NPAR K-S(NORMAL) = extra neur sexism.
```

to test whether the frequencies of the values for any of these (Gender Attitudes Survey) computed variables (*not* recoded into categories) depart significantly from a normal distribution (with mean and standard deviation unspecified – and therefore taken from the observed frequency distributions).

Two-sample test

The two-sample Kolmogorov-Smirnov test compares the distribution of frequencies for two observed groups defined by an **independent** (grouping) variable. Thus it could be used to compare the frequency distribution of answers to a seven-point scale question for, say, males and females. The data used should not be categorical (i.e. it should be at least at the ordinal level).

Such a command would take the following form:

```
SPSS/PC: NPAR K-S = var001 BY sex(1,2).
```

To compare the *sex* distribution for a number of variables a list could be used:

```
SPSS/PC: NPAR K-S = var001 to var003 BY sex(1,2).
```

This would give an analysis of the similarity of the distribution for the two sexes for each of the three variables specified. A non-significant probability would indicate that the distributions of the answers for the two sex groups do not differ significantly.

The display includes the count of valid cases, the **K-S** *Z* value (with its two-tailed probability) and the most extreme differences between the two groups.

13.4 Two-sample tests

The Wald-Wolfowitz test – NPAR W-W

The Wald-Wolfowitz test (W-W) analyzes the scores for a dependent variable from two **independent** samples. It rank-orders all of the scores together, from low values to high values, as if they came from a single population. A runs test analysis is then performed on the ranks in order to determine whether the dependent variable is distributed evenly throughout the ranked sequence. If the two samples differ substantially in their central tendency, their variability or their skewness, then a significant Z value will be obtained.

Thus if boys (B) and girls (G) of a certain age were rank-ordered for height, for example, the sequence:

```
BGGBGBGBBBGGGBBGGGBBBBGBGBGBBBGGBG
```

would suggest that height was not strongly associated with sex (i.e. heights were not different between the sexes). The sequence contains 20 runs. However, the sequence:

```
GGGGBGGGGGBGGGGGBGBBBBGGBBBBBBGBBB
```

might suggest that boys were taller than girls (there are 12 runs).

Each display includes the count for each group, the number of runs (or the minimum and maximum number of runs if there are ties between the groups). A Z value is given for the number of runs, together with the one-tailed probability.

Suppose that we have the idea that males make more facial movements or hand gestures when watching violent films than females. We could count the number of such movements (the variable *moves*) as groups of men and women watch a fifteen-minute presentation of violent film extracts. An appropriate SPSS/PC+ command for analyzing the data with a Wald-Wolfowitz test would be:

```
SPSS/PC: NPAR W-W = moves BY sex(1,2).
```

The Moses test – NPAR Moses

The 'Moses test of extreme reactions' (for two **independent** samples) examines whether the **range** of an ordinal variable is similar for the two groups. One of the groups acts as the control group, and the other is the comparison group.

The Moses test is useful when we suspect that one group of subjects has extreme reactions (i.e. both low and high values) compared to another group.

Thus the two groups might have similar means or similar mean ranks, but their 'span' might be significantly different. Data from both groups is rank ordered in a single ascending sequence and the span of the control group is found. From the size of the total sequence the probability of such a span occurring by chance can be calculated. To override distortions which can arise from presence of the odd extreme value in the control group, a degree of trimming can be implemented. By default the extreme 5 per cent of cases from each end (known as 'outliers') are trimmed in this way.

The display includes the counts of valid cases in the two groups, the span of the control group (with and without the outliers) and the associated one-tailed probabilities.

Suppose we believe that when passing a poster proclaiming the health dangers of smoking, smokers are likely to exhibit 'extreme' behaviours (compared to non-smokers), either passing very quickly by or pausing to examine the contents in detail. We could measure the time taken to pass by the poster (*passtime*) for both smokers (coded 1 on the variable *smoker*) and non-smokers (coded 0) and then subject our hypothesis to the Moses test with the following SPSS/PC+ command:

```
SPSS/PC: NPAR Moses = passtime BY smoker(0,1).
```

The order of the values specified for the variable *smoker* determines which group is treated as the control group and which is treated as the experimental group. The first value specifies the control group (in this case non-smokers, value 0). We might decide to trim 10% of 'control group outliers' from either end of the rank sequence. With, say, 200 non-smokers as controls we would do this by specifying that 20 non-smokers should be trimmed from both ends, using the following command:

```
SPSS/PC: NPAR MOSES(20) = passtime BY smoker(0,1).
```

The Mann-Whitney test – NPAR M-W

The Mann-Whitney (M-W) test is a two-sample test which compares the ranks of two **independent** groups defined by a 'grouping' variable. (It is therefore in many ways similar to the independent *t*-test procedure for parametric data.) The values of the dependent variable are rank-ordered for both groups as if they came from a single population. The test statistic U is calculated in a formula which includes $n1$ and $n2$ (the numbers in the two groups) and the larger of the two totals of ranks for the groups. The *lower* the value of U the more significant are the ranks of the two conditions (groups).

The output display includes, for both groups, the valid count and the mean rank of the variable. If there are fewer than 30 cases then the U statistic is given with its exact two-tailed probability. If there are more cases then U is converted to a *Z* statistic. The Wilcoxon *W* value is also given (see later), together with its probability.

The **M-W** subcommand could be used, for example, to determine whether there was a significant sex difference in the number of calories which subjects said they consumed in an average day – the variable *cals*.

A suitable command would be:

```
SPSS/PC: NPAR M-W = cals BY sex(1,2).
```

This command would cause the system to rank order cases in terms of their value for *cals* and then to sum the ranks for each of the conditions (the male condition and the female condition). The higher sum of ranks would then be used to calculate the value of *U*.

The McNemar test – NPAR Mcnemar

The McNemar Test is used to compare two dichotomous **related** variables which have been coded with the same two values. If the variables to be tested are not already in this form then the data should be **RECODED** and the recoded variables used for **NPAR Mcnemar**. The test is particularly useful for before-and-after designs where one variable represents a dichotomy before an event or manipulation and the other variable represents the same dichotomy after the event or manipulation. The data can be nominal.

Suppose that we wish to know whether voting intentions (in a two-party state) have changed significantly since a particular event (for example, a scandal involving a prominent politician). If we have a record of subjects' voting intentions *before* the scandal broke (*vote1*), *immediately after* the scandal broke (*vote2*), and some *three weeks after* the scandal broke (*vote3*) then we can analyze the changes which have taken place in voting intentions.

The display presents a 2 X 2 table with variable names (or labels) and values, the total number of valid cases and the two-tailed binomial probability.

The simplest form of the command would be:

```
SPSS/PC: NPAR Mcnemar = vote1 vote2 vote3.
```

This would produce 3 separate McNemar tests, because each *pair* of variables included in the variable list would be tested together (in this case *vote1* with *vote2*, *vote1* with *vote3*, and *vote2* with *vote3*).

The keyword **WITH** can be used. Thus:

```
SPSS/PC: NPAR Mcnemar = vote1 WITH vote2 vote3.
```

would produce just *two* tests (*vote1* with *vote2*, and *vote1* with *vote3*).

A variable list can be included on either side, or both sides, of the keyword **WITH** (see also the section below on **Options**).

A McNemar test can be used to compare the 'ideal date' and 'ideal marriage partner' data (*dat01* to *dat08* and *mar01* to *mar08*) from the Gender Attitude Survey. These data can be recoded so that the 'very important' rating retains the value 1 and all other responses take the value 2:

```
SPSS/PC: recode dat01 to dat08 mar01 to mar08 (2,3,4 =2).
```

The McNemar test can now be used to see whether the frequency of the 'very important' rating is different for ideal dating and marriage partners for each of the eight attributes included:

```
SPSS/PC: NPAR Mcnemar dat01 to dat08 with mar01 to mar08
       : /option = 3.
```

Option 3 is specified here to request a special **WITH** pairing (as for the **T-TEST**). This means that *dat01* will be compared only with *mar01*, that *dat02* will be compared only with *mar02*, etc.

The output from this command included the following:

```
- - - - - McNemar Test

     DAT08        Dependability (date)
with MAR08        Dependability (spouse)

                            MAR08

                              2             1             Cases      43
                  |_____________|_____________|
           1      |           1 |        17   |
DAT08             |_____________|_____________|  (Binomial)
           2      |          15 |        10   |  2-tailed P   .0117
                  |_____________|_____________|
```

(1 is 'very important'; other ratings are represented by 2.) Here we see a significant difference in the number of respondents stressing dependability as a 'very important' characteristic of a date and of a spouse. Ten respondents felt that dependability was a 'very important' characteristic of an ideal spouse but not of an ideal date. For only one subject was the reverse true.

The sign test – NPAR Sign

The sign test examines pairs of data (i.e. it compares the values of two specified variables for each case) to test whether the number of positive differences between the variables (variable A greater than variable B) is the same as the number of negative differences (variable B greater than variable A). Clearly such comparisons mean that data must be able to be ordered, so data subjected to the sign test must be at least at the ordinal level of measurement.

The display provides counts of the number of positive and negative differences, the number of ties (i.e. no difference between the two values), and the two-tailed binomial probability of the sign distribution obtained.

Suppose we are interested in whether subjects who have previously taken a general knowledge test (without being told whether the answers they gave were right or wrong) are as likely to get a lower score when tested again as they are to get a higher score. If the two scores obtained provide the values for the variables *test* and *retest*, the following **Sign** procedure could be used to determine whether subjects were equally likely to get a higher or lower score on retest:

```
SPSS/PC: NPAR SIGN = test retest.
```

The keyword **WITH** can also be used, with variable lists on either or both sides of the keyword (this is requested with **Option 3** – for further details of the **NPAR Options** see Section 13.7).

The Wilcoxon test – NPAR Wilcoxon

The Wilcoxon test (or Wilcoxon signed-ranks test) is used to assess whether two **related** samples (with data at the ordinal level or above) are significantly different (it is in many ways equivalent to the matched-pairs *t*-test procedure for parametric data). The differences between the pairs of variables are ranked for magnitude, irrespective of whether the differences are positive or negative. The sums of the ranks are then determined separately for those cases in which the difference is positive and those cases in which the difference is negative, and the smaller of these sums is the value W. From the number of pairs and the value of W, another statistic, Z, can be calculated and its probability determined.

The **NPAR Wilcoxon** output display includes the total valid count and the mean rank for both variables. The number of positive and negative rank differences and the number of tied ranks is also given. The Z value is included together with its two-tailed probability.

Suppose that we wish to test the effectiveness of a drug which is claimed to improve memory performance. We could test subjects' normal memory (*normem*, without the drug) and then the same subjects' memory after they have been given the drug (*drugmem*). We wish to answer the question, 'Is there a significant difference between the paired values (one pair per case) of the two variables *normem* and *drugmen*?'

The following **NPAR** command could be used to determine whether there was such a difference:

```
SPSS/PC: NPAR Wilcoxon = normem drugmem.
```

The keyword **WITH** can also be used, with variable lists on either or both sides of the keyword. (For further details on specifying the comparisons to be made see Section 13.7 on **Options**).

13.5 Two- or *k*-sample tests

The median test – NPAR Median

The median test is a test for **independent** groups for which data are available at the ordinal level of measurement. It can be used to compare either two groups (the **two sample median test**) or more than two groups (the **extended median test**), and determines whether the groups differ with respect to a median value. For both versions of the test the data are cast into a 2 X *k* crosstabulation, where each of the two rows represents a split of values at the overall median (or some other specified cutting point) and each of the *k* columns represents a particular group.

The two-sample median test
The overall median (for the two groups combined) is calculated and used to divide the values for each group into two. A 2 X 2 table is formed and appears in the display. The two cells in the first column of this display contain the frequencies for values above the median and below (or at) the median. The second column provides the corresponding values for the second group. From this table a chi-square statistic is calculated. The display includes the crosstabulation, the number of valid cases, the overall median, and the chi-square statistic and its significance (probability).

Suppose that we have asked a group of subjects to list their phobias (fears of spiders, snakes, heights, confined spaces, etc.). The data for the variable *fears* (the number of phobias each person has listed) may be highly skewed, with many subjects reporting few fears and some reporting many. To test whether there is a significant difference in the number of fears reported by men and by women we could use the median test. The variables used would be *fears* and *sex*.

An appropriate SPSS/PC+ **NPAR** command for requesting this two-sample median test would be:

```
SPSS/PC: NPAR Median = fears BY sex(1,2).
```

Here the cutting point would be the overall median of the two groups combined. Alternatively, we could specify a cutting point:

```
SPSS/PC: NPAR Median (3) = fears BY sex(1,2).
```

Cases with more than three fears would be assigned to the top row of the crosstabulation. Cases with three or fewer fears would be assigned to the bottom row.

The extended median test

The median test can also be used to determine whether there are overall differences for more than two groups (*k* groups). The data is cast into a 2 X *k* crosstabulation, with the rows again representing a split of values at the overall median (or some other specified cutting point) and each of the *k* columns representing a particular group. A chi-square statistic is calculated from this table, and the display includes the number of degrees of freedom for the table as well as the information provided for the two-sample test.

Suppose that, on the basis of their scores on a personality questionnaire, subjects are each assigned to one of three anxiety level groups (high anxiety, medium anxiety and low anxiety – coded 1, 2 and 3 respectively on the variable *anxiety*). If we now wish to determine whether there are differences between these groups in terms of the number of specific fears or phobias they have previously reported (the variable *fears*), we could use the following SPSS/PC+ command.

```
SPSS/PC: NPAR Median = fears BY anxiety(1,3).
```

This would give a chi-square statistic for a 2 X 3 table in which each of the three groups was included, with a split at the overall median.

On the other hand, reversing the parameters specified for the variable *anxiety*:

```
SPSS/PC: NPAR Median = fears BY anxiety(3,1).
```

would yield a two-sample median test with only groups 3 and 1 included.

The Kendall test – NPAR Kendall

The Kendall test available within the SPSS/PC+ **NPAR** procedure calculates *W*, the coefficient of concordance. This is a measure of the degree to which a number of judges (two or more) agree in their rating (or ranking) of a number of items. *W* has a range from 0 to 1, where 0 means that there is *no* agreement between the judges and 1 means that there is *total* agreement. In more general terms, the Kendall test assesses whether two or more (*k*) *related* samples could be from the same population.

The display includes the mean rank (across all judges) for each variable, the total number of judges (here 'cases'), *W*, the chi-square equivalent, degrees of freedom (= number of variables – 1) and the significance value.

Suppose that 5 judges have been called upon to rate 12 photographs (variables *photo1* to *photo12*) on a 10-point scale of, say, attractiveness. To assess the overall agreement between the judges we could use the following SPSS/PC+ **NPAR** command:

```
SPSS/PC: NPAR Kendall = photo1 to photo12.
```

13.6 *k*-sample tests

The Kruskal-Wallis test – NPAR K-W

The Kruskal-Wallis (K-W) test is a non-parametric one-way analysis of variance and is similar to the Mann-Whitney (except that three or more conditions may be involved). The test is used to establish whether there is a significant overall difference in a dependent variable for a number of groups differentiated on a particular independent variable. It is therefore an **independent** samples test. The data must be at least ordinal. The data are ranked (all as one population) and the rank-sum for each group is calculated. The measure obtained is *H*.

The display includes the mean rank and the count for each group, the number of valid cases, the chi-square equivalent of *H*, and its probability value (significance). Where there are tied ranks across groups the chi-square value corrected for the ties is also displayed, again with the significance.

Suppose that we wish to discover whether there are differences in the attitude towards capital punishment (*attcp*) between groups of subjects who have been given one of three relevant reports to read. One report is heavily biased in favour of capital punishment, one is dispassionate and essentially neutral, and the other is highly critical. The variable *report* is coded according to which report has been read and the three reports have been assigned the values 1, 2 and 3. *Report* is the independent variable and *attcp* the dependent variable.

The **NPAR** command used to request a Kruskal-Wallis one-way analysis of variance for this study could be:

```
SPSS/PC: NPAR K-W = attcp BY report(1,3).
```

This command creates three *report* groups (the numbers in brackets indicate the minimum and maximum values to be used in forming the groups). The dependent variable *attcp* is ranked across all groups and the sums of the ranks for each group are then compared.

The Cochran test – NPAR Cochran

The Cochran test can be regarded as an extended form of the McNemar test. It assesses whether, for several **related** samples, the proportions assigned to a particular (dichotomous) value (nominal or ordinal) are significantly different. Thus if, in a series of test questions, *similar* proportions of subjects supply the correct answer to each question, the Cochran test statistic – *Q* – would *not* be significant. If, however, there were substantial differences between these proportions, then *Q would* be significant.

The test involves the construction of a 2 X *k* crosstabulation (where each of the two rows contains the count for one or other side of the dichotomy, and *k* is the number of variables).

The display gives the count for each of the two values for each variable, the total valid count, and the Cochran *Q* statistic with the appropriate degrees of freedom and significance value.

Suppose that we have respondents' answers to eight general knowledge questions (*Q1* to *Q8*) and we wish to establish whether the difficulty level of the questions is approximately equal. The answers could be scored as 'right' or 'wrong' and the Cochran test used to determine whether, overall, there is any significant difference in the difficulty between questions.

A suitable SPSS/PC+ command for requesting such a Cochran analysis would be:

```
SPSS/PC: NPAR Cochran = Q1 to Q8.
```

The Friedman test – NPAR Friedman

The Friedman test can be regarded as an extension of the Wilcoxon test. It is a two-way analysis of variance by ranks which tests whether it is likely that two or more *related* samples (variables) could have come from the same population. If the chi-square value produced is significant then it can be concluded that the variables have *not* come from the same population.

Data from a number (*n*) of cases are available for each of a number (*k*) of conditions. The data must be at least at the ordinal level of measurement. The data are cast into a *n* X *k* table. In the Friedman analysis, the data from each case are ranked *horizontally* from 1 to *k* and a two-way analysis of variance (cases by conditions) is performed. If some conditions (variables) have a substantially lower (or higher) sum of ranks than others then this indicates that the conditions are significantly different.

The **NPAR Friedman** output display includes the mean rank for each of the variables (conditions), the degrees of freedom for the table, and the chi-square value and associated significance.

Suppose that five children have each been asked to rank-order eight flavours of ice-cream (the variables *ice1* to *ice8*) in terms of their own preference. The data matrix in Table 13.2 has emerged (1 means 'ranked first' and 8 means 'ranked last').

Table 13.2

Child	Ice-cream Flavours (*k*)
	1 2 3 4 5 6 7 8
001	8 6 1 3 5 7 2 4
002	7 5 3 2 6 8 1 4
003	6 7 5 1 8 4 2 3
004	8 4 2 3 7 6 1 5
005	8 7 2 3 6 5 1 4

The data given in the body of the table are the actual ranks as chosen by the children. If they had 'rated' the ice creams instead (e.g. given them points from 1 to 20) then SPSS/PC+ would rank the data before performing the Friedman test.

To obtain a Friedman analysis, the appropriate command would be:

```
SPSS/PC: NPAR Friedman = ice1 to ice8.
```

With the above data the following display was obtained:

```
- - - - - Friedman Two-way ANOVA


    Mean Rank   Variable        Cases          Chi-square
        7.40     ICE1            5               28.2000
        5.80     ICE2
    ...                          D.F.      Significance
        6.00     ICE6              7               .0002
        1.40     ICE7
        4.00     ICE8
```

13.7 NPAR: Statistics and Options

Additional statistics may be requested. In many cases, however, only some of these additional statistics will be valid or meaningful.

Statistic 1 requests that the mean, maximum, minimum, standard deviation and count of each variable should be displayed.

Statistic 2 requests the display of the count and of quartiles (i.e. the values which provide the cutting points for 25 per cent, 50 per cent and 75 per cent of the frequency distribution for each variable).

By default, variables with missing values are deleted from analyses of single samples. For two or more samples they are deleted pairwise.

Option 1 requests that missing values should be included in the analysis.

Option 2 requests the listwise deletion of cases with missing values. A case with a missing value for any variable listed on a subcommand is excluded from all analyses requested by that subcommand.

Option 3 concerns the pairing of variables in subcommands requesting analysis with two *related* samples (it can therefore be used only with three **NPAR** tests – McNemar, sign and Wilcoxon). This option may be specified with or without the keyword **WITH**. There are thus four possibilities, as shown in Table 13.3.

Table 13.3

WITH included?	Option 3 specified?	
No	No	**Every variable paired with every other**
Yes	No	**Each variable to left of WITH paired with each variable to the right of WITH**
No	Yes	**Sequential pairs tested (i.e. 1 with 2, 3 with 4, 5 with 6, etc.)**
Yes	Yes	**1st var before WITH paired only with 1st after, 2nd with 2nd, etc.**

Option 4 applies only when the data file to be analyzed is too large for the available computer memory. If such a situation occurs, and this option has been specified, then a random sample of the data is selected and analyzed.

13.8 The RANK procedure

Several of the non-parametric analyses available through **NPAR** work with ranks. They rank-order the data before performing the test, so that the data submitted for analysis can be entered in either interval or ordinal form.

Even when data are available at the interval level, it is often useful to rank-order them. Sometimes we are more interested in the ordinal value (the position of a data item within an ordered list) than in an absolute value. Thus in the case of a league table of sports performance, it is common to quote a team's position in the league rather than the number of games they have won or the number of points they have scored.

In such cases, ranking is based on points awarded, goals scored, games won, time taken, or another such measure. In a *descending* ranking, rank 1 (the top rank) will be assigned to the case (team, athlete, etc.) with the highest points score. Rank 2 will be assigned to the case with the next highest score, and so on. In an *ascending* ranking, however, rank 1 will be assigned to the case with the lowest measure, rank 2 to the case with the next lowest measure, and so on. Thus if the time to complete a race is recorded in seconds, rank 1 might be assigned to the athlete who completed the course in the least number of seconds (i.e. the fastest person), rank 2 to the next 'fastest', and so on.

This is a straightforward practice until we meet cases with equal scores. There are then a variety of ways in which 'tied ranks' can be assigned.

A number of useful analyses can be performed using the SPSS/PC+ **RANK** procedure. **RANK** ranks any specified variable and creates a new variable in the active file, the values of which correspond to either an ascending rank-ordering or a descending rank-ordering of the original variable.

The command:

```
SPSS/PC: RANK salary(d).
```

creates a new variable *rsalary*, a rank order score for each case in the population based on descending order of the original *salary* score. The case with the highest score for *salary* now has a *rsalary* score of 1, the highest has a *rsalary* score of 2, etc. The descending order was specified by the (d) that followed the variable name.

Without this specification, the new variable *rsalary* would have reflected an ascending rank order (i.e. the lowest salary would have been ranked 1, etc.).

By default, the name of the variable created is the original variable name prefaced with an r, the variable is given the label "RANK of (ORIGINAL)", and the names for the original and the new variable are displayed, together with the new variable name, in the following way:

```
From      New
variable  variable  Label
--------  --------  -----

SALARY    RSALARY   RANK of SALARY
```

Values for the new variable can be displayed by a **LIST** command. For example:

```
SPSS/PC: LIST id rsalary.
```

This will produce a listing which includes the values for the variable *id* (subject identification number) and for *rsalary*.

An alternative name for a ranked variable to be created can be specified by using an additional **Rank** *sub*command. For example:

```
SPSS/PC: RANK salary / Rank into ranksal.
```

The variable created will now have the name *ranksal*.

By default, a summary table showing the name of any variable which has been ranked, together with the name of the new variable, is written to the listing file.

Missing cases are normally excluded from **RANK** operations, although the user can specify that missing values should be included (and therefore ranked as if they were real values) by a **Missing** subcommand used in the following way:

```
SPSS/PC: RANK salary / Rank into ranksal / Missing = include.
```

System missing values are always assigned system missing values for any variable created by the **RANK** command.

Ranking within groups

Sometimes it is more useful to rank order a particular variable within specific groups that make up the population. Thus a variable might be ranked separately for males and females. To rank the variable *salary* for cases within each faculty group (the variable *fac*), for example, we would use the following command:

```
SPSS/PC: RANK salary(d) by fac.
```

If we wished to list the results of this procedure, the output would be more easily understandable if we sorted the cases in the active file by *fac* before issuing a **LIST** command. The output display would then list all cases for *fac* value 1 ('arts' in the Gender Attitudes Survey), then all cases for value 2 ('science'), etc. The following sequence of commands could be used to create the appropriate rank variable (within groups), to specify the name *rksalfac* for this variable, to sort cases in the file by faculty and finally to produce a display ordered by faculty:

```
SPSS/PC: RANK salary(d) by fac/ Rank into rksalfac.
SPSS/PC: SORT by fac.
SPSS/PC: LIST id fac rksalfac.
```

Part of the display produced by this set of commands is shown below:

```
ID FAC  RKSALFAC

 1   1    13.000
 2   1    14.000
 4   1     5.500
 5   1    10.500
 ....
20   1     7.500
29   1    16.500
35   1     3.000
 3   2     1.000
12   2     6.000
```

```
15   2     3.000
24   2    10.000
27   2    12.000
....
45   2    10.000
46   2     6.000
11   3    12.500
21   3    17.500
22   3     7.000
26   3     1.500
....
48   3    20.000
49   3    12.500
50   3     4.000
```

The Ties subcommand

It was mentioned above that there are a number of alternative ranking strategies for dealing with equal scores. 'Tied ranks' can be assigned in a number of different ways. The **Ties** subcommand specifies which of four methods is to be used.

The four strategies can be illustrated by the use of the following set of scores, one for each of eight cases A to H:

Case	A	B	C	D	E	F	G	H
Score	14	7	14	9	12	13	20	18

First let us arrange the set in a descending sequence (the rationale would be precisely the same, throughout, if we had chosen an ascending order instead). Thus:

Case	G	H	A	C	F	E	D	B
Score	20	18	14	14	13	12	9	7

*The **Mean** strategy*

Ranking in descending order, there is no problem in assigning the rank '1' to the score of 20, and the rank '2' to the score of 18. If every score were unique the next two ranks would be '3' and '4', but in fact the next two scores are both 14. Using the **Mean** strategy, we would assign the average rank to both scores (i.e. 3.5). Thus the rankings for the set would be:

Case	G	H	A	C	F	E	D	B
Rank	1	2	3.5	3.5	5	6	7	8

The ***Low*** *strategy*

An alternative strategy is to give each of a set of (two or more) tied scores the next available rank. Thus in the example both scores of 14 would be given the rank of '3'. However, the next score, 13, would be assigned the rank of '5' (*not* '4'), and continuing for the remainder of the data, the lowest score (7) would receive the last rank '8' (and there are, of course, eight items within the set):

Case	G	H	A	C	F	E	D	B
Rank	1	2	3	3	5	6	7	8

The ***High*** *strategy*

A variation on this strategy is to give each of a set of (two or more) tied scores the highest available rank. Thus in the example both scores of 14 would be given the rank of '4' (the rank '3' would not be assigned):

Case	G	H	A	C	F	E	D	B
Rank	1	2	4	4	5	6	7	8

The ***Condense*** *strategy*

Finally, a set of (two or more) tied scores might be given the next available rank ('3' in the case of our two scores of 14) but in this case the next score, 13, would be assigned the rank of '4' (*not* '5'), and continuing for the remainder of the data, the lowest score (7) would receive the last rank '7' (despite the fact that there are eight items within the set):

Case	G	H	A	C	F	E	D	B
Rank	1	2	3	3	4	5	6	7

The **Ties** subcommand takes the following form:

```
SPSS/PC: RANK sexism(d) / Ties = low.
```

If no **Ties** subcommand is included, then the default strategy (Mean) is used to deal with tied ranks.

Rank-order correlation

After two or more variables have been ranked for the whole population, specifying the newly created (rank) variables within a **CORRELATION** command requests a Pearson correlation. The result is a Spearman Rank Correlation Coefficient.

For example:

```
SPSS/PC: RANK sexism  neur.
SPSS/PC: CORRELATION rsexism rneur.
```

Notice that an ascending rank-ordering has been requested (by default) for both of the variables *sexism* and *neur.* A positive correlation would therefore tell us that high ranking cases of *rsexism* tend also to have high ranks for *rneur.* However, the value of the correlation is in fact 0.0236 and is therefore non-significant.

The Ntiles subcommand

A number of other subcommands are available for **RANK**. The **Rank** subcommand, which we have already met, is one of a number of 'rank function subcommands'. Another of these is **Ntiles**, which creates a new variable by first grouping ranked values into a specified number of percentile groups. The new variable has values from 1 to *n*, where *n* is the number of specified groups. Thus to divide cases into quartiles on the basis of their height in inches (*ht*), the following command could be used:

```
SPSS/PC: RANK ht/ Ntiles(4).
```

This would create a new variable which, by default, would be named *nht*, because an n is prefixed to the original variable name when **Ntiles** is requested. The name can be changed, however, by the use of 'INTO' within the subcommand. Thus:

```
SPSS/PC: RANK ht/ Ntiles(4) into quartht.
```

would lead to the creation of a quartile variable named *quartht*. This variable would have four values (1 to 4). Cases which were within the first quartile of the population (i.e. the quarter of cases with the lowest height) would have the value 1 for *quartht*. Cases in the second quartile would have the value 2 for *quartht*, and so on.

Clearly, it would be useful to supply both a variable label and value labels for this new variable. For example:

```
SPSS/PC: RANK ht/ Ntiles(4) into quartht.
SPSS/PC: VARIABLE LABEL quartht "Height by (ascending) quartile".
SPSS/PC: VALUE LABELS quartht 1 "Height 1st quartile - short"
         : 2 "Height 2nd quartile" 1 "Height 3rd quartile"
         : 4 "Height 4th quartile - tall".
```

A basic **FREQUENCIES** command for *quartht* produced the following output:

```
 QUARTHT   Height by (ascending) quartile

                                                    Valid     Cum
Value Label              Value  Frequency  Percent Percent  Percent

Height 1st quartile          1         11     22.0    23.4     23.4
Height 2nd quartile          2         11     22.0    23.4     46.8
Height 3rd quartile          3         15     30.0    31.9     78.7
Height 4th quartile          4         10     20.0    21.3    100.0
                             .          3      6.0 Missing
                                  -------  -------------
                         Total         50    100.0   100.0

Valid cases        47      Missing cases       3
```

Notice that the value labels provided were truncated in the output display, so that the words 'short' and 'tall' do not appear.

14 *Statistical Analysis VI: RELIABILITY*

14.1 The concept of reliability

It is often claimed that a particular set of items (such as a series of closed-ended questions related to a particular topic), can be used together as a complex 'measure' of a single variable such as intelligence, an aspect of personality, or a person's attitude to a particular issue.

How can we tell if such a measure is any good? Two questions are vital. Is the measure reliable? And if so, is it valid? Both reliability and validity are complex concepts. Taking validity first, we can say that a test or measure is valid if it provides a true measure of what it is purported to be measuring. Thus a test of intelligence is valid if it actually measures intelligence.

A reliable measure provides consistent and stable measurements. If a test that is claimed to provide a measure of some stable characteristic of an individual produces one score on one occasion and a very different score on another occasion, for example, then it cannot be said to be a reliable measure. And since it offers a variety of scores, of which only one at most can be valid, then the limited reliability places limits on the validity of test scores. (It must be added that if a test is highly reliable it must be measuring *something*, but there is no guarantee that it is measuring what it is supposed to measure, i.e. that it is valid.)

The 'observed score' on a test reflects the 'true' score to some degree, and 'error' to some degree. Reliable scores are those for which the error component is low and the 'true score' component is relatively high, and reliability can be defined as the ratio of true variance to observed variance. Thus the aim in estimating the reliability of a test is to assess how much of the variability in test scores is due to error and how much it reflects the variability in the true scores.

14.2 The assessment of reliability

Four strategies are commonly used to assess reliability.

The test-retest method

The test-retest strategy involves administering the same test twice to the same group of cases and then correlating the first set of scores with the second set of scores. There are several problems with this method. It requires two administrations, separated by time, and the length of separation time is likely to affect the reliability coefficient obtained. Also, the characteristic being measured might change over this time period, so that the 'true score' will have changed. Furthermore, people who complete a test for the second time may be affected by their previous experience with the test. They might find it more boring on the second occasion, for example, or they might remember some of the previous responses they gave.

Alternate forms method

Another strategy used to assess reliability involves the creation of two forms of a test which are closely parallel in terms of length, style and content. By giving subjects both tests, and then correlating the scores obtained from both forms, an estimate of the test reliability is obtained. This strategy avoids several of the problems of the rest-retest strategy but does leave open the possibility that completing the first form may affect responses to the second form in some way. There is also nothing to guarantee that the two forms of the test are strictly parallel.

Split-half methods

When data have been collected for a test composed of several items, it is possible to create, in effect, two alternate forms by splitting the test into two halves and correlating the scores obtained from the two halves. Splits may be achieved in various ways, for example by splitting a scale into 'the first half' and 'the second half', or by separating the 'odd' and the 'even' items, or by taking a random selection of half of the items and then assigning the remaining items to the other half. However, for long tests there are a very large number of possible splits. Thus a ten-item test can be split into two sets of five items in 126 different ways, and each split might yield a different estimate of the reliability of the test.

Internal consistency methods

Whereas split-half methods compare scores based on halves of the test, internal consistency methods compare each item to every other item. The most widely used measure of internal consistency, the alpha coefficient, is mathematically equivalent to the average of the reliability coefficients that would be obtained from all possible split-halves.

Internal consistency methods estimate the reliability of a test on the basis of the number of items in the test and the average correlations between pairs of items. Cronbach's 'alpha' is an estimate of the correlation to be expected between scores obtained on our scale (the 'observed' score) and a 'true score' (in effect, the score that would be obtained from a perfect measure of the scale variable). The value of alpha depends on the number of items that make up the scale and the correlations between them. The greater the number of items, and the greater the correlations between the items, the higher the value of alpha. Alpha values vary between -1 and +1, as do other correlation coefficients, but a negative score would clearly indicate that the collection of items should not be included within the same scale. The higher the alpha value the better (more reliable) the scale.

It is possible to calculate the value of alpha if any particular item were to be deleted from the scale. If such information shows that the alpha value would increase substantially following the deletion of an item, this might lead us to revise our scale, leaving out the item that is reducing the alpha value and weakening the reliability of the scale as a whole.

14.3 The RELIABILITY procedure

In Chapter 8 we showed how new variables could be created from existing variables by using a **COMPUTE** command. In one example, scores reflecting levels of agreement with a number of statements about gender differences were added together to form a *sexism* score for each respondent. However, although the statements were all related to gender attitudes, the addition of scores to form a single scale was based on the questionable assumption that each of the item scores was measuring the same thing (a variable that we named "sexism").

The **RELIABILITY** procedure will allow us to explore the statistical relationships between data relating to different items, so that we can determine whether a particular set can be taken together as a measurement of a single dimension, i.e. whether the composite score reflects a reliable **scale**.

The SPSS/PC+ **RELIABILITY** procedure is used in the following way. A list of variables is specified (on a **Variables** subcommand) and then one or more **Scale** subcommands are used to specify the scales that are to be analyzed. Each **Scale** subcommand provides a name for the scale (in brackets) and lists the variables, the values for which are to added together to form the corresponding scale ('TO' can be used to specify consecutive variables). The specification 'ALL' means that all of the variables specified on the original **Variables** subcommand should be added together to form the scale.

However, **RELIABILITY** does not create a new variable (i.e. the scale variable) in the active file. To create such a variable, after a **RELIABILITY** analysis, you should use an appropriate **COMPUTE** command.

The Model subcommand

Five different types of model can be used within a **RELIABILITY** command to assess the reliability of a scale. We will describe the use of only two of these, the internal consistency model (requested by 'alpha' on a **Model** subcommand), and the 'split-half' model (requested on a **Model** subcommand by 'split').

Alpha is the default model used by the SPSS/PC+ **RELIABILITY** procedure, and it is therefore used if the **Model** subcommand is omitted. Alternatively you can specify the model by including the subcommand:

```
/ Model = alpha
```

When a split-half model is requested within a SPSS/PC+ **RELIABILITY** procedure the score obtained from the first half of the items is compared with the score obtained from the second half of the items (if there are an odd number of items, the first half is larger by one item). A **Model** subcommand to specify that the split-half method for assessing reliability should be used, would take the following form:

```
/ Model = split
```

However, it is possible to override the default division into two halves by specifying how many items should be included in the *second* half:

```
/ Model = split(12)
```

If the above subcommand related to a scale composed of 36 items, the split would therefore be made with 24 items in the first 'half' and 12 in the second.

The Statistics subcommand

The **Statistics** subcommand is used to request various additional statistics (by name).

DESC requests that certain descriptive statistics (means and standard deviations) be displayed for each item.
CORRELATIONS requests that the matrix of correlations between each pair of items be displayed.
SCALE requests the mean value and variance for the scale (or for each scale, if more than one **Scale** subcommand is included within the **RELIABILITY** command).

Six other statistics specifications are available.

The Summary subcommand

The **Summary** subcommand is used to request summary statistics for the items in the scale.

MEANS takes the mean values of every item in the scale as the dataset and then displays statistics relating to this set of mean values, including the average (i.e. the overall mean), the minimum mean value, the maximum mean value, the range of means, the ratio of the minimum mean to the maximum mean, and the variance of the means.

VARIANCE displays the same set of statistics as MEANS, but takes as the dataset the variance for every item in the scale.

CORRELATIONS again displays the same set of summary statistics, but in this case they refer not to means or variances but to the correlations between every pair of scale items.

TOTAL produces statistics that compare each item with the scale composed of all *other* items. It therefore allows us to see what effect there would be if the current item were deleted from the scale. Thus for a five-item scale, the statistics displayed beside item 1 would show various characteristics of a scale composed only of items 2, 3, 4, and 5. The statistics displayed beside item 2 show various characteristics of a scale composed only of items 1, 3, 4, and 5, and so on. The statistics displayed when TOTAL is requested include the scale mean, the scale variance, and Cronbach's alpha (all based on a scale that *excludes* the specified item). Another piece of information displayed, and one which is particularly important for scale construction, is the correlation between each item and the scale composed of all other items.

14.4 The use of RELIABILITY: an example using 'alpha'

To illustrate the use of the **RELIABILITY** procedure we will employ variables from the Gender Attitudes Survey, and attempt to construct a reliable measure of 'sexism' from the variables *var001* to *var010*. We will do this by testing the *sexism* scale we developed in Chapter 8. You may recall that the variables *var001* to *var010* relate to 10 attitude statements from the Gender Attitude Questionnaire. Respondents were asked to indicate how far they agreed or disagreed with each statement by circling one number on a 7-point scale from 1 to 7, where 1 was labelled 'Strongly Agree' and 7 was labelled 'Strongly disagree'.

In the earlier Chapter it was suggested that statements 1, 2 and 9 were 'non-sexist', implying that for these items a *high* score (indicating *disagreement*) would indicate a *high* degree of *sexism*. On the other hand, statements 3, 4, 5 and 8 were judged to be 'sexist', implying that *agreement* with these statements (indicated by a *low* coded value) would reflect a sexist attitude. Thus for these items a low value was made to *increase* the *sexism* score significantly.

The three remaining statements of the original 10 were judged to be somewhat ambiguous with regard to their 'sexist' or 'non-sexist' nature and they were therefore not included in the attempt to derive a composite *sexism* score.

The *sexism* score for each respondent was obtained by combining information from both the 'positive sexism' items and the 'negative sexism' items by adding together the scores obtained for each of the 'positive sexism' items and adding the *reversed* scores for the 'negative sexism' items. Reversing the score meant that a 7 became 1, a 6 became 2, etc. The general formula used to reverse the scores was '(8 – original score) = reversed score'.

A single *sexism* score for each respondent was specified within a single **COMPUTE** command. Thus:

```
SPSS/PC: compute sexism = (var001 + var002 + var009 +
       : (8 - var003) + (8 - var004) + (8 - var005) +
       : (8 - var008)).
```

This way of obtaining values for *sexism* is hardly satisfactory. It relies too heavily on untested assumptions about how the items relate to one another, and the assumption that they can be added to form a meaningful and useful (reliable) scale is certainly open to challenge.

We will now use the **RELIABILITY** procedure to test out some of the relevant assumptions and to establish whether or not the above formula really does provide a satisfactory scale.

Since it is not appropriate for present purposes to use a **COMPUTE** statement to derive a composite *sexism* score, we need to work with the original seven variables that were entered into the above formula. But before we do anything else we must reverse the coding for four of these statements, so that for every item a high score indicates 'high sexism' (**RELIABILITY** always makes a scale by *adding* all of the values for the specified items).

We could reverse the scores for the relevant variables with a **RECODE** command. Thus:

```
SPSS/PC: RECODE var003 to var005 var 008
       : (7 = 1) (6 = 2) (5 = 3) (3 = 5) (2 = 6) (1 = 7).
```

or we might wish to keep the original variables intact and therefore create additional 'reversed' variables which we can use in the **RELIABILITY** procedure. We could do this with the following **COMPUTE** commands:

```
     /PC: compute var003r = (8 - var003).
SPSS/PC: compute var004r = (8 - var004).
SPSS/PC: compute var005r = (8 - var005).
SPSS/PC: compute var008r = (8 - var008).
```

Thus the seven variables to be entered into our scale are *var001*, *var002*, *var009*, *var003r*, *var004r*, *var005r* and *var008r*.

To explore the reliability characteristics of the *ad hoc* sexism scale previously used, the following **RELIABILITY** command was issued:

```
SPSS/PC: RELIABILITY / variables = var001 var002 var009
       : var003r var004r var005r var008r
       : / scale (sexism) = all
       : / statistics = desc correlations scale
       : / summary = means variances correlations total.
```

The **RELIABILITY** command here includes four subcommands. **Variables** specifies which variables are to be used in the analysis. **Scale** provides a name for a scale ('sexism') and specifies which of the variables listed on the **Variables** subcommand are to be included in this scale (in this case 'ALL'). Several **Scale** commands, each with a different subset of the variables previously listed on the **Variables** subcommand, may be included with a single **RELIABILITY** command.

The **Statistics** subcommand requests means and standard deviations for each item (DESC), a matrix of correlations between each pair of items (CORRELATIONS), and the mean and variance for the scale (SCALE).

Finally, the **Summary** subcommand requests summary statistics relating to the MEANS and VARIANCES for each item in the scale, and to the CORRELATIONS between each pair of items in the scale. In addition, specifying TOTAL will produce statistics comparing each individual item with a scale composed of all *other* items. Thus values will be presented for the mean, variance and alpha value of a scale that *excludes*, in succession, each specific item. TOTAL will also display a value for the correlation between each item and the scale composed of all other items.

The output from the above command extends over several pages. Let us examine it, one step at a time.

```
R E L I A B I L I T Y   A N A L Y S I S  -  S C A L E  (S E X I S M)

1.     VAR001      Women equal to men
2.     VAR002      New legislation needed
3.     VAR009      Men less marriage commitment
4.     VAR003R
5.     VAR004R
6.     VAR005R
7.     VAR008R
```

This is a key to the variables that constitute the scale to be examined. Variable labels are displayed for the three variables that have been provided with labels (the newly created variables were not given labels).

		MEAN	STD DEV	CASES
1.	VAR001	3.0833	1.7238	48.0
2.	VAR002	2.1250	1.6711	48.0
3.	VAR009	5.0417	1.4286	48.0
4.	VAR003R	3.4583	1.6626	48.0
5.	VAR004R	4.0625	1.7553	48.0
6.	VAR005R	3.2292	1.5877	48.0
7.	VAR008R	3.2292	1.9045	48.0

As requested on the **Statistics** subcommand (by 'DESC'), the means and standard deviations for each item are displayed. A count of the number of cases on which these statistics are based is also provided. The **Statistics** subcommand also requested 'CORRELATIONS', and the correlations between each pair of items is displayed in the next part of the output:

CORRELATION MATRIX

	VAR001	VAR002	VAR009	VAR003R	VAR004R
VAR001	1.0000				
VAR002	.3656	1.0000			
VAR009	-.2088	-.1894	1.0000		
VAR003R	.1720	.1857	-.4024	1.0000	
VAR004R	.3850	.2657	-.2726	.2451	1.0000
VAR005R	.0084	.0531	-.1825	.0883	.1780
VAR008R	.2662	.2850	-.2617	.3894	.3520

	VAR005R	VAR008R
VAR005R	1.0000	
VAR008R	.3693	1.0000

The fact that most of the correlations are low indicates that they are not measuring the same thing, and the fact that some of the correlations are negative is particularly disturbing. However, we see that all of the negative correlations involve *var009*, and this suggests that this item is a 'maverick'. This variable relates to the statement **'Most men make a much less serious commitment to marriage than women'**. In the original (*ad hoc*) formula used to calculate an overall *sexism* score, *dis*agreement with this statement was treated as 'sexist', but even our simple correlational analysis indicates that this assumption was wrong. Indeed, if anything it appears that *agreement* with the statement indicates a 'sexist' point of view.

The **Statistics** subcommand also requested 'SCALE', and the next part of the output provides various statistics for the scale:

```
       N OF CASES =                   48.0
                                                   N OF
STATISTICS FOR          MEAN   VARIANCE    STD DEV VARIABLES
    SCALE            24.2292  34.1804     5.8464       7
```

From this display we see that 48 cases have been analyzed and that the scale is composed of seven variables. The average score obtained by respondents on the scale as it stands is 24.23 and the variance is 34.23. The standard deviation of the scores is equal to the square root of the variance, and is therefore 5.85.

As part of the same display table, some of the information requested by the **Summary** subcommand is displayed.

```
ITEM MEANS    MEAN   MINIMUM   MAXIMUM   RANGE     MAX/MIN   VARIANCE
            3.4613   2.1250    5.0417    2.9167    2.3725       .8159

ITEM
VARIANCES     MEAN   MINIMUM   MAXIMUM   RANGE     MAX/MIN   VARIANCE
            2.8283   2.0408    3.6272    1.5864    1.7774       .2405

INTER-ITEM
CORRS.        MEAN   MINIMUM   MAXIMUM   RANGE     MAX/MIN   VARIANCE
             .0996   -.4024     .3894     .7918    -.9679       .0630
```

The 'ITEM MEANS' row of information is based on the mean values for each of the seven items in the scale. Thus the value given in the first column tells us that the 'mean of the 7 mean values' is 3.46, the value in the second column tells us that the 'minimum of the 7 mean values' is 2.13, and so on. The range of means (highest mean minus the lowest mean) is included in the display, as are the ratio of the highest mean to the lowest mean ('max/min') and the 'variance of the 7 mean values'. The second row of information provides the corresponding statistics for the variances of the 7 items. The third and final row provides similar information, but the values displayed do not relate to data on particular items but to values for the correlations between pairs of items. The effect of the 'maverick' variable *var009* is again apparent, for we see that the minimum correlation (-0.40; we see from the previously displayed correlation matrix that this is the correlation between *var009* and *var003r*) has a greater absolute value (ignoring the – sign) than the highest positive correlation.

The final specification on the **Summary** subcommand was 'TOTAL', and this produced the following table:

ITEM-TOTAL STATISTICS

	SCALE MEAN IF ITEM DELETED	SCALE VARIANCE IF ITEM DELETED	CORRECTED ITEM-TOTAL CORR.	SQUARED MULTIPLE CORR.	ALPHA IF ITEM DELETED
VAR001	21.1458	25.0208	.3588	.2443	.3930
VAR002	22.1042	25.4996	.3489	.1848	.3997
VAR009	19.1875	39.6024	-.4151	.2197	.6619
VAR003R	20.7708	26.9464	.2590	.2607	.4414
VAR004R	20.1667	23.8014	.4261	.2485	.3572
VAR005R	21.0000	28.4681	.1884	.1716	.4717
VAR008R	21.0000	21.3617	.5221	.3444	.2916

The first column of values shows what the average score for the scale would be if the 'row variable' were omitted from the scale. Information provided in a previous table indicated that the scale mean for the 7-item scale is 24.2292, and the first value in the present table tells us that the mean would be reduced to 21.1458 if *var001* were deleted from the scale. Working down the first value column, we see that the mean for a six-item scale omitting *var002* would be 22.1042, and so on. The second column provides similar information, except that this time the values indicate what the variance of the scale would be if the row item were deleted. The 7-item scale variance was 34.1804 (shown in a previous table), and we can therefore conclude that the omission of most items would reduce the variance. However, once again *var009* proves an exception, because the table shows that the omission of this variable would actually increase the variance. By reducing the variance, *var009* is having the effect of 'levelling out' the scores, and this may be considered undesirable because we want the scale to differentiate between individuals as far as possible.

The next column provides the 'corrected item-total correlation'. This is the correlation between values for the row item and a score derived from a scale omitting that item. This is therefore a good measure of how well each item corresponds to the mass of other items. As we might expect by now, the value for *var009* is strongly negative, suggesting that, if anything, it is measuring the *opposite* of what the other items are measuring.

The 'squared multiple correlation' column gives the value for a statistic (R^2) calculated from a multiple regression equation. In this equation the row item is the 'dependent variable', and every other item is acts as an independent variable. The question is how well the value for the row item can be predicted from an equation involving the values for all other variables in the scale. To put it another way, the issue concerns how much of the variability in the row item can be explained by the other items. The values obtained are rather low, indicating that scores for the set of 'other' scale items would not be very useful to us in an attempt to predict the score of one particular item from the scale. The low values obtained for R^2 are hardly surprising, given that the values for the correlations in the inter-item correlation matrix were also generally low.

The final column, 'Alpha if item deleted', provides us with values for the Cronbach alpha reliability coefficient for scales that exclude one specific item. These coefficients will be more meaningful if we examine them in comparison to the overall alpha value for the full seven-item scale, which is provided (as default) in the final part of the output from the **RELIABILITY** command:

```
RELIABILITY COEFFICIENTS      7 ITEMS

ALPHA =    .4909     STANDARDIZED ITEM ALPHA = .4364
```

From this we see that the alpha value for the whole scale is .49. The 'standardized item alpha' gives the value for alpha if scores on all of the items were transformed to 'standard form' so that the variance for each item was 1.

If we compare the alpha value for the full 7-item scale (.49) with the alpha values that would result from deleting individual items from the scale (as shown in the previous table) we find that deleting any item would reduce the reliability of the scale, except in the case of *var009*. With *var009* deleted, the alpha value would increase sharply.

Thus the **RELIABILITY** analysis has shown us, through many of the statistics produced, that *var009* does not belong in this scale. We would therefore wish to delete the item from our scale. The alpha value for the 6 item scale would then be .66. This is not a high value (it is generally recommended that alphas of .80 and above are 'acceptable' for a scale).

In fact, rather than deleting the item, we might consider including it, but reversing the value of the scores, because the fact that negative correlation coefficients were found between this item and all other items in the scale suggests that a reversed score might add to the reliability of the scale. If we do this (using a **COMPUTE** command to create a new variable *var009r* by reversing the scoring for *var009*) then we do obtain a slightly higher alpha value (.70) for the new 7-item scale.

Although the estimated reliability of even the revised scale is considerably below the value we would expect of a major test of personality or intelligence, for example, it does reflect a moderate to low reliability. The reliability coefficient obtained suggests that the 'true score' is contributing rather more to the observed score than error is contributing. The low reliability is also partly a function of the number of items that make up the scale. Seven items make a very short scale, and increasing the number of 'appropriate' items would certainly increase the reliability of the scale (provided that the average correlation between the items does not change).

14.5 The use of RELIABILITY: an example using 'Split'

If we revise the **RELIABILITY** command (with seven items, including the newly 'reversed' variable *var009r*), but this time specifying that the split-half model should be used (by specifying 'split' on the **Model** subcommand), we obtain the following:

```
SPSS/PC: RELIABILITY / variables = var001 var002
       : var003r var004r var005r var008r var009r
       : / scale (sexism) = all
       : / model = split
       : / statistics = desc correlations scale
       : / summary = means variances correlations total.
```

Two things have changed since we submitted the original command. First, the variable list has been changed, to include *var009r* rather than *var009*. Secondly, the **Model** subcommand has changed (previously it was left out, so that the default 'alpha' model was used).

One effect of changing the variable list can be seen in the correlation matrix which forms part of the output from the above command. All of the negative correlations have been replaced by the positive correlations of precisely the same absolute value (as we would expect following the transformation used to create *var009r* from *var009*).

CORRELATION MATRIX

	VAR001	VAR002	VAR009R	VAR003R	VAR004R
VAR001	1.0000				
VAR002	.3656	1.0000			
VAR009R	.2088	.1894	1.0000		
VAR003R	.1720	.1857	.4024	1.0000	
VAR004R	.3850	.2657	.2726	.2451	1.0000
VAR005R	.0084	.0531	.1825	.0883	.1780
VAR008R	.2662	.2850	.2617	.3894	.3520

	VAR005R	VAR008R
VAR005R	1.0000	
VAR008R	.3693	1.0000

The **Statistics** subcommand which specified 'SCALE' provides a new set of statistics for the scale as a whole, and also for 'Part 1' (the first half of the scale, consisting of just four items) and for 'PART 2' (the second half of the scale, consisting of just three items):

```
          N OF CASES =              48.0
                                                          N OF
STATISTICS FOR         MEAN    VARIANCE     STD DEV   VARIABLES
      PART 1        11.6250     18.5372      4.3055           4
      PART 2        10.5208     14.8081      3.8481           3
      SCALE         22.1458     49.1059      7.0076           7

ITEM MEANS      MEAN   MINIMUM   MAXIMUM   RANGE     MAX/MIN  VARIANCE
     PART 1   2.9063    2.1250    3.4583   1.3333     1.6275     .3164
     PART 2   3.5069    3.2292    4.0625    .8333     1.2581     .2315
     SCALE    3.1637    2.1250    4.0625   1.9375     1.9118     .3385

ITEM VARIANCES MEAN   MINIMUM   MAXIMUM   RANGE     MAX/MIN  VARIANCE
     PART 1   2.6423    2.0408    2.9716    .9309     1.4561     .1692
     PART 2   3.0764    2.5208    3.6272   1.1064     1.4389     .3060
     SCALE    2.8283    2.0408    3.6272   1.5864     1.7774     .2405

INTER-ITEM
CORRELATIONS    MEAN   MINIMUM   MAXIMUM   RANGE     MAX/MIN  VARIANCE
     PART 1    .2540     .1720     .4024    .2304     2.3396     .0095
     PART 2    .2998     .1780     .3693    .1913     2.0747     .0090
     SCALE     .2441     .0084     .4024    .3940    47.7786     .0122
```

The item-total statistics are also provided:

```
ITEM-TOTAL STATISTICS

             SCALE       SCALE    CORRECTED
             MEAN      VARIANCE     ITEM-       SQUARED        ALPHA
            IF ITEM     IF ITEM     TOTAL      MULTIPLE       IF ITEM
            DELETED     DELETED     CORR.        CORR.        DELETED

VAR001      19.0625     37.8896     .3885        .2443         .6671
VAR002      20.0208     38.6166     .3706        .1848         .6716
VAR009R     19.1875     39.6024     .4151        .2197         .6619
VAR003R     18.6875     38.0492     .4043        .2607         .6628
VAR004R     18.0833     35.9929     .4763        .2485         .6427
VAR005R     18.9167     41.7376     .2363        .1716         .7033
VAR008R     18.9167     33.4397     .5466        .3444         .6197
```

And finally, the split-half reliability coefficients are displayed:

```
RELIABILITY COEFFICIENTS      7 ITEMS

CORRELATION BETWEEN FORMS =          .4756

EQUAL LENGTH SPEARMAN-BROWN =        .6446

GUTTMAN SPLIT-HALF =                 .6419

UNEQUAL-LENGTH SPEARMAN-BROWN =      .6481

ALPHA FOR PART 1 =    .5731     ALPHA FOR PART 2 =    .5651

4 ITEMS IN PART 1               3 ITEMS IN PART 2
```

The 'correlation between forms' is the correlation of the two scores for each of the 48 cases, one score based on the four-item scale and one score based on the three-item scale.

The 'equal length Spearman-Brown' value is an estimate of the reliability of the whole test, but is based on the assumption that both of the split halves are of the same length, an assumption that does not hold in the present case. Instead, we need to refer to the 'unequal length Spearman-Brown' coefficient, because the two halves of the scale consist of an unequal number of items, 4 and 3. The Spearman-Brown correlations assume that the two halves of the test are equally reliable and have similar variance. These assumptions are not made in the case of the 'Guttman split-half' coefficient, which also provides an estimate of the reliability of the test as a whole.

Finally, alpha values are given for each part of the test.

In the example we have used, the best estimate of the reliability of the whole test is provided either by the Guttman split-half coefficient (0.64) or by the unequal-length Spearman-Brown coefficient (0.65). We can see that the values for these two coefficients are very similar, and that both are slightly below the value for alpha (0.70) established earlier.

We can therefore conclude that our scale of seven items has provided *some* measure of a common variable, although the reliability estimates are not high.

14.6 Using the scale

There is therefore some justification in adding the scores obtained for the relevant variables and computing a score for a new variable. However, we must be careful not to assume too readily that we know the nature of this common variable. Interpretation of the meaning of a scale is a particular skill, and hypotheses may be disconfirmed when the validity question is investigated. Thus if a scale that we are assuming to be a measure does not correlate with established measures of that construct, we would need to think again about what our scale is measuring.

Having provided a name for a computed variable, it is very easy to assume thereafter that the name really does identify the construct being measured. But the variable name is no more than a user-supplied tag, and care should be taken that it is not allowed to assume an unwarranted characterization as a 'real thing'.

For the example developed in the previous section, the creation of the new variable (with the name *sexism2*) could be achieved, using original and newly created (reverse coding) variables, with the following **COMPUTE** command:

```
SPSS/PC: COMPUTE sexism2 = var001 + var002 + var003r
       : + var004r + var005r + var008r + var009r.
```

We know that this variable has greater reliability than the original *sexism* variable, and it would therefore be of some interest to see how the newly derived *sexism2* variable relates to other variables in the original dataset. An independent (groups) *t*-test was performed to determine whether there was a significant difference in *sexism2* scores between the males and the females in the sample. The mean *sexism2* score for males is found to be 25.57, whereas the mean score for female respondents is 19.48. A *t*-value of 3.28 was obtained, the corresponding probability being .002. Thus there is a highly significant difference between male and female *sexism2* scores, with males having significantly higher scores. However, as explained above, we would have to exercise care in interpreting this to mean that the males in the sample were more 'sexist' than the females.

15 Joining Files

Sometimes, after a file has been created, new data are obtained and the user wishes to add them to the original file. New **variables** or new **cases** may need to be added.

New variables: Suppose that a file contains the results of a political attitude survey taken before an election, and that information has subsequently been collected from many of the same subjects *after* the election. If the user wishes to analyze changes and to examine relationships between some of the original and some of the new variables, all of the relevant variables must be contained within a single file. A composite file therefore needs to be created, containing both the old and the new **variables** (the maximum number of variables that can be included in any active file or system file is 500).

New cases: Another situation might arise. Suppose that a large number of questionnaires have been sent out in a postal survey. After the first batch of questionnaires has been returned a data file may be created, but for some time afterwards further returns may continue to arrive. Eventually a final summary and analysis of the results will be needed for all of the cases together. A composite file will need to be created, containing both the old and the new **cases**.

One simple and direct way of expanding an existing file to add new cases or new variables would be to return to the original data file and to insert the additional information with an editor program (such as **REVIEW**), amending the data definition file as appropriate. This is one of the advantages of continuing to work with an ASCII file (in this case a data file) rather than with a system file.

However, if you have your data in a system file then the new data (i.e. the new cases, or new variables for the original cases) could be entered into a second file and a **JOIN** command then used to create a new composite file. This would include both the information from the original file and that from the second file.

The **JOIN ADD** command is used to add new **cases**; the **JOIN MATCH** command is used to add new **variables**.

Many of the same rules apply to the **JOIN MATCH** and **JOIN ADD** commands:

- up to five input files can be joined to form a single composite file (there is a system limit of 500 variables in any file).
- the files to be joined must be **system** files (except that one can be the current active file – the active file is specified in the **JOIN** command by an asterisk *)
- unless a special 'order subcommand' (**BY**) is in effect, the order of the cases or variables in the composite file is determined by the order in which the input files are named on the command.

Several **subcommands** can also be used within **JOIN** commands:

- the user can **Drop** specified variables from the files to be joined
- the user can **Keep** specified variables from the files to be joined
- the user can **Rename** variables before they are entered into the new composite file
- the user can issue a **Map** subcommand to identify the source of variables in the composite file (and the effects of any **Rename** instructions)

The **LIST** command can be used to list the data in the new file, and the **SAVE** command can be used to save the contents of the new file to disk.

15.1 The JOIN MATCH command

The **JOIN MATCH** command is used to create a new file containing the **variables** from two files (for overlapping cases). Suppose that the simple data definition file:

```
DATA LIST FILE= "file1.dat" / id 1-3 var001 to var010 4-13.
```

refers to a data file ("file1.dat") which includes 10 cases and contains the following data:

```
0011111111111
0021111111111
0031111111111
0041111111111
0051111111111
0061111111111
0071111111111
0081111111111
0091111111111
0101111111111
```

To **JOIN** the data from this file with those from another file we would first create a system file (e.g. "file1.sys") containing the data. To create such a file from the data definition file ("file1.def"), we would issue the command:

```
SPSS/PC: INCLUDE "file1.def".
```

and then create a system file by using the **SAVE** command. Thus:

```
SPSS/PC: save outfile= "file1.sys".
```

Imagine that further data has been collected for the same subjects. The new data definition file ("file2.def") is:

```
DATA LIST FILE= "file2.dat" / id 1-3 var011 to var020 4-13.
```

and the new data file ("file2.dat") contains the following data:

```
0012222222222
0022222222222
0032222222222
0042222222222
0052222222222
0062222222222
0072222222222
0082222222222
0092222222222
0102222222222
```

Again, we would create a system file ("file2.sys") from the data-definition file. With versions of both of the relevant files as system files we can now create a composite file with a **JOIN** command:

```
SPSS/PC: JOIN MATCH FILE= "file1.sys"
       : /FILE= "file2.sys".
SPSS/PC: LIST.
```

(If "file2.def" had been used to create the current active file then the second subcommand above would be replaced with '/FILE = *.').

Although the variable *id* is included in both of the original files, it will occur only once in the composite file. This is because the variable has the same *name* in both of the original files. If different names had been used, the composite file would include the subject identifier twice, under different names.

The order of the filenames on the **JOIN MATCH** command ("file1.sys" followed by "file2.sys") determines the order in which the variables will appear on the composite file.

The above command would produce an active file composed of material from both of the source files. The **LIST** command would produce the following display:

ID	VAR001	VAR002	VAR003	VAR004	VAR005	VAR006	VAR007	VAR008	VAR009	VAR010	VAR011	VAR012	VAR013	VAR014	VAR015	VAR016	VAR017	VAR018	VAR019	VAR020
1	1	1	1	1	1	1	1	1	1	1	2	2	2	2	2	2	2	2	2	2
2	1	1	1	1	1	1	1	1	1	1	2	2	2	2	2	2	2	2	2	2
3	1	1	1	1	1	1	1	1	1	1	2	2	2	2	2	2	2	2	2	2
4	1	1	1	1	1	1	1	1	1	1	2	2	2	2	2	2	2	2	2	2
5	1	1	1	1	1	1	1	1	1	1	2	2	2	2	2	2	2	2	2	2
6	1	1	1	1	1	1	1	1	1	1	2	2	2	2	2	2	2	2	2	2
7	1	1	1	1	1	1	1	1	1	1	2	2	2	2	2	2	2	2	2	2
8	1	1	1	1	1	1	1	1	1	1	2	2	2	2	2	2	2	2	2	2
9	1	1	1	1	1	1	1	1	1	1	2	2	2	2	2	2	2	2	2	2
10	1	1	1	1	1	1	1	1	1	1	2	2	2	2	2	2	2	2	2	2

The composite file has now become the active file. It can be **SAVED** as a system file by using a command such as:

```
SPSS/PC: SAVE OUTFILE= "compfile.sys".
```

The two input files which were joined in the above example ("file1.sys" and "file2.sys") are said to be **parallel** because each contains data for the same cases in exactly the same order. The second file contains extra variables for *every* case in the first file. There are no missing cases and no extra cases.

In real studies such a situation would be rare. Commonly, follow-up data is obtained from only *some* of the original cases. Suppose that the second data file contained data from only 7 of the original 10 cases, respondents 4, 6 and 9 having failed to reply to a follow-up questionnaire. Unless the **JOIN MATCH** command above were modified, the system would assign the data from the second phase of the study to subjects 1 to 7. The system would treat the remaining subjects (8 through 10) as if they had not provided follow-up data (it would assign system-missing values to these **cases** for each of the follow-up variables). Clearly this would provide a serious mis-matching which would produce major errors in subsequent analyses.

Such mis-matching can be prevented by providing instructions on how the data in the second file should be assigned to cases from the first file. The most efficient way of doing this is to include a subject *id* variable in *each* of the files to be joined, and then to instruct the program to match **BY** *id*. The **BY** subcommand must follow after all other file specifications. Thus:

```
SPSS/PC: JOIN MATCH FILE= "file1.sys"
       : /FILE= "file2.sys"
       : /BY id.
SPSS/PC: LIST.
```

This would produce a composite file containing some system-missing values, and these would be indicated by a period (.) in any subsequent **LIST** display. Thus with no second-phase data available for subjects 4, 6 and 9, the **LIST** command would produce the following output:

```
     V V V V V V V V V V V V V V V V V V V V
     A A A A A A A A A A A A A A A A A A A A
     R R R R R R R R R R R R R R R R R R R R
     0 0 0 0 0 0 0 0 0 0 0 0 0 0 0 0 0 0 0 0
     0 0 0 0 0 0 0 0 0 1 1 1 1 1 1 1 1 1 1 2
ID   1 2 3 4 5 6 7 8 9 0 1 2 3 4 5 6 7 8 9 0

 1   1 1 1 1 1 1 1 1 1 1 2 2 2 2 2 2 2 2 2 2
 2   1 1 1 1 1 1 1 1 1 1 2 2 2 2 2 2 2 2 2 2
 3   1 1 1 1 1 1 1 1 1 1 2 2 2 2 2 2 2 2 2 2
 4   1 1 1 1 1 1 1 1 1 1 . . . . . . . . . .
 5   1 1 1 1 1 1 1 1 1 1 2 2 2 2 2 2 2 2 2 2
 6   1 1 1 1 1 1 1 1 1 1 . . . . . . . . . .
 7   1 1 1 1 1 1 1 1 1 1 2 2 2 2 2 2 2 2 2 2
 8   1 1 1 1 1 1 1 1 1 1 2 2 2 2 2 2 2 2 2 2
 9   1 1 1 1 1 1 1 1 1 1 . . . . . . . . . .
10   1 1 1 1 1 1 1 1 1 1 2 2 2 2 2 2 2 2 2 2
```

Each of the input files to be included in the composite file must already be sorted in the ascending order of any variable used (by being specified after the **BY** keyword) to bring about a match. Complex matching procedures can be specified (with up to 10 **BY** keywords used within a single command), but the simplest way of arranging a match is to include the *id* variable for each respondent on each of the input files. If the data are then entered (or re-sorted) in ascending *id* order, the files can be matched using the **BY** subcommand ("BY *id*").

The Drop and Keep subcommands

Variables can be dropped from the composite file by specifying **Drop**, followed by a variable list, after the relevant input file name. Thus:

```
SPSS/PC: JOIN MATCH FILE= "file1.sys"
       : /DROP=var004
       : /FILE= "file2.sys"
       : /DROP= var011 var012 var013 var018
       : /BY id.
SPSS/PC: LIST.
```

If only a few of the variables from a particular file are to be entered into the composite file it may be more efficient to specify which ones should be retained. In this case a **Keep** subcommand would be used. This should specify, after the relevant input file name, the list of the variables to be included in the composite file.

N.B. The TO convention is NOT permitted when variables are listed on Drop or Keep subcommands within a JOIN command.

The Rename subcommand

Variables can be renamed before they enter the composite by specifying **Rename** and listing the changes to be made in the following manner:

```
SPSS/PC: JOIN MATCH FILE= "file1.sys"
       : /RENAME (var001 var003 var005 = att1 att3 att5)
       : /DROP=var004
       : /FILE= "file2.sys"
       : /RENAME (var014 = item3)
       : /DROP=var011 var012 var013 var018
       : /BY id.
SPSS/PC: LIST.
```

Notice that in the **Rename** subcommand either a single variable or a number of variables may be renamed within a pair of brackets with a = sign between the old and the new variables, thus:

```
(old1 = new1)   or   (old1 old2 = new1 new2)
```

The TO convention is NOT permitted on the Rename subcommand of the JOIN command.

The above command, with the **Drop** and **Rename** subcommands as specified, produced the listing shown opposite. Here, the *id* variable and the next nine variables have come from the input file "file1.sys". One of the variables from this source file has been dropped and three have been renamed. The remaining six variables come from "file2.sys". Four of the original 10 variables in this file have been excluded as a result of the **Drop** subcommand and one (*id*) is not featured because it is a duplication of the *id* variable from the first named file. One of the variables has been renamed

ID	ATT1	VAR002	ATT3	ATT5	VAR006	VAR007	VAR008	VAR009	VAR010	ITEM3	VAR015	VAR016	VAR017	VAR019	VAR020
1	1	1	1	1	1	1	1	1	1	2	2	2	2	2	2
2	1	1	1	1	1	1	1	1	1	2	2	2	2	2	2
3	1	1	1	1	1	1	1	1	1	2	2	2	2	2	2
4	1	1	1	1	1	1	1	1	1	.	.	.	.	.	.
5	1	1	1	1	1	1	1	1	1	2	2	2	2	2	2
6	1	1	1	1	1	1	1	1	1	.	.	.	.	.	.
7	1	1	1	1	1	1	1	1	1	2	2	2	2	2	2
8	1	1	1	1	1	1	1	1	1	2	2	2	2	2	2
9	1	1	1	1	1	1	1	1	1	.	.	.	.	.	.
10	1	1	1	1	1	1	1	1	1	2	2	2	2	2	2

The file "file2.sys" contains only seven cases, and system-missing values have been assigned where **cases** are missing. The files have been matched using the **BY** keyword.

The contents of the composite file

If variable labels, value labels and missing values are specified on any of the input files then the composite file will retain any of the information relevant to the data it contains. If two or more files contain *different* definitional information regarding particular data, then the composite file will include the version which was *first* encountered in the specified sequence of input files.

Similarly, if the same variable **name** is encountered in more than one of the input files, the value entered into the composite file will be the *first* value encountered (even if this is a missing value).

15.2 The JOIN ADD command

Suppose that we have created a file of the initial returns from a postal survey and we now wish to add the data from later returns. We thus have new **cases** with the same variables as the original cases.

If we have retained the original data file, then adding new cases is a very simple matter. Typically, the new cases would first be entered into a new file, and when this file was complete and had been checked for typing or coding errors it would be added to the end of the original data file. This would be done by editing the original file with REVIEW. With the cursor on the bottom line of the file, pressing **<F3><i>** would give the instruction to insert another file. Entering the name of the file containing data for the new cases, in response to the prompt, the file being edited would now contain both the original and the new data. The complete file would then be saved, retaining the original name, by entering **<F9><w>** to save the whole file and then pressing <RETURN> to accept the (original) file name offered after the prompt.

If the original data is now in the form of a system file, however, the data from the early and the late returns would be combined by creating a new file with the **JOIN ADD** command.

As with **JOIN MATCH**, the input files to be combined into a composite file must be system files (except that one can be the active file). The simplest form of the **JOIN ADD** command is:

```
SPSS/PC: JOIN ADD FILE= "data1.sys"
       : /FILE= "data2.sys".
```

This would create a new (active) file which would contain the cases from "data1.sys" followed by the cases from "data2.sys".

Thus if the first input file contains 100 cases, and the second input file contains 56 cases, the composite file will contain 156 cases. Each case in the composite file will contain all of the variables included on *any* of the input files (system-missing values will be provided if a file does not contain a variable which is included in one of the other input files).

Suppose that 500 questionnaires, pre-coded 001 to 500, have been mailed. When some of the completed questionnaires have been returned an initial data file may be created. Rather than sorting the questionnaires physically into the code order we could simply work through the pile of returns, entering first the code number (as the variable *id*), followed by the other data. The first data file might contain data from 200 questionnaires. The order of the cases in that file, in terms of the *id* numbers, will be haphazard or random.

At this stage a data definition file would be created, and we may wish to conduct some preliminary analyses. During this process the decision may be made (perhaps because we need the data processing to be a speedy as possible) to work with a system file.

After a further month we may decide that the data collection is now at an end. Perhaps another 150 questionnaires will have been received, making 350 (or a 70% return) altogether. Data from the second batch can be entered into a second data file, again in haphazard *id* order.

If the second set of data is to be added to the first (using **JOIN ADD**), it will first need to be defined by a version of the original data definition file, using exactly the same variable names. The file can then be made active, and either joined directly to the existing system file, or saved as a system file before being joined. The two system files would then be joined by a **JOIN ADD** command, e.g.:

```
SPSS/PC: JOIN ADD FILE= "data1.sys"
       : /FILE= "data2.sys".
```

The composite file could then be re-ordered with a **SORT** command so that cases appear in their original *id* order.

```
SPSS/PC: SORT BY id.
```

The re-ordered active file could then be saved:

```
SPSS/PC: SAVE OUTFILE= "combined.sys".
```

As with **JOIN MATCH**, up to five files can be joined with **JOIN ADD**.

The subcommands **Rename**, **Drop** and **Keep** can be used to select and name the **variables** within the composite file (**cases** *cannot* be dropped). **Rename Keep** and **Drop** subcommands apply only to the file named immediately before, and will therefore need to be repeated if the same variables are to renamed, kept or dropped from different source files. Thus:

```
SPSS/PC: JOIN ADD FILE= "data1.sys"
        : /RENAME= (var001 var003 var005 = item1 item3 item5)
        : /DROP= var002 var004 var006 var007 var008 var009
        : /FILE= "data2.sys"
        : /RENAME= (var001 var003 var005 = item1 item3 item5)
        : /DROP= var002 var004 var006 var007 var008 var009
        : /BY id  /map.
```

Notice that each of the variables to be dropped is specifically named. The **TO** convention is *not* permitted on a **Rename**, **Keep** or **Drop** subcommand of a **JOIN** command.

The **Map** subcommand can be used at any stage within the command to list the variables on the active file together with their source. If **Map** is included as the final subcommand it will show the final state of the active file (with the heading 'RESULT'). The **Map** request within the above command produced the following output:

RESULT	DATA1.SYS	DATA2.SYS
ID	ID	ID
ITEM1	VAR001	VAR001
ITEM3	VAR003	VAR003
ITEM5	VAR005	VAR005
VAR010	VAR010	VAR010

16 *Advanced Uses of the System*

This chapter concerns three topics:

- some further ways in which data can be entered into SPSS/PC+ files
- how **portable files** may be created and transported between the SPSS/PC+ microcomputer and other computers (including mainframe computers running SPSS^{X})
- some commands from the SPSS/PC+ base system which have not been included in this presentation, and some of the facilities offered by the various optional modules that are available.

16.1 New data options

By this time the *SPSS/PC+ 4.0 Base Manual* is likely to have become an indispensable aid, and you should now have become familiar enough with the SPSS/PC+ system to be able to understand much of the information it provides.

This book will have enabled you to explore a dataset, and to carry out a range of analyses, and you should now have a good grasp of the way in which the SPSS/PC+ system works. The presentation has avoided many of the ifs and buts which would have been necessary to provide a truly comprehensive account, but you may now feel sufficiently skilled and confident to explore further ways of using the system.

In this section we will consider some alternative ways in which data may be entered. The datasets we have been dealing with throughout this book have consisted entirely of whole numbers, without a decimal point, and the data for each variable have been assigned to specific columns within a rectangular dataset. SPSS/PC+, however, offers considerably more flexibility than this. First of all, the data within a dataset do not all have to be in number form. SPSS/PC+ also allows data to be entered in character form. Thus alphanumeric data, which might include names of people, or of countries, for example, may be represented within the dataset by character **strings**. Secondly, the strategy we have followed in assigning particular columns in the dataset to particular variables (**fixed format**) is actually one of two ways in which data can be arranged (the other is **free format**). And whereas we have been assuming that all of our data are in the form of integer (whole) numbers, SPSS/PC+ can handle decimal data (i.e. data which include digits to the right of a decimal point) perfectly well. Furthermore, the system provides a facility for entering decimal data in such a way that the decimal point is *assumed*. We will discuss these additional facilities in turn.

Strings

Throughout this book, all of the data have been represented in the form of numerals whether the values have been truly numerical or categorical. But SPSS/PC+ can also deal with alphanumeric data such as the names and addresses of respondents.

Thus a data file ("a.dat") could look like this:

```
001  Mary      Dobson  21433123145645621
002  John      Croft   12433123145645621
003  Robert    Crystal 11433123145645621
004  Susan     Lewin   22433123145645621
```

A data list ("a.def") for the above file might be:

```
DATA LIST FILE = "a.dat"/ id 1-3 forename 6-12 (A)
surname 16-23 (A) sex 24 agegroup 25 var001 to var015 26-40.
```

The (**A**) specification which follows the column allocation for the variables *forename* and *surname* indicate that these two variables are alphanumeric **strings**.

Strings which are eight characters long or less are referred to as 'short strings'. Other strings (which may be up to 255 characters in length) are referred to as 'long strings'. Each eight-character portion of a long string counts as one variable towards the system limit of 500 variables (thus a nine-character string would count as two variables, a 20-character string as three variables, etc.). Certain manipulations are possible with short strings but not with long strings – for example, only short strings can be used in transformation commands (e.g. **COMPUTE** and **RECODE** commands).

When a string is specified in a command it must be enclosed within quotation marks, and cannot be split across two lines.

A **short** string variable can be created with an **IF** command. In the following example:

```
SPSS/PC: if (sex eq 1) sexalph = "  male".
SPSS/PC: if (sex eq 2) sexalph = "female".
```

male cases would be assigned the value 'male' for the new alphanumeric variable *sexalph*, and the value 'female' would be assigned to this variable for female subjects.

If short strings are included in transformation commands, the specification must be **precise**. Leading or trailing blanks must be included and lower-case and upper-case characters are *not* treated as equivalent. Strings cannot be compared to numbers, or to numerical variables. The following set of commands was used to create a new variable *group* from an original numerical variable *agegroup* and the newly created short string variable *sexalph*:

```
SPSS/PC: if (sexalph eq "  male" and agegroup eq 1) group = "  BOY".
SPSS/PC: if (sexalph eq "  male" and agegroup eq 2) group = "  MAN".
SPSS/PC: if (sexalph eq "female" and agegroup eq 1) group = " GIRL".
SPSS/PC: if (sexalph eq "female" and agegroup eq 2) group = " WOMAN".
```

Used with the data from the file "a.dat", and after *sexalph* had been computed, this set of commands created the new short string alphanumeric variable *group*. The following frequencies display was then produced in response to the command '**FREQUENCIES group.**':

```
GROUP
                                                Valid     Cum
 Value Label         Value  Frequency  Percent  Percent  Percent

                       BOY          1     25.0     25.0     25.0
                       MAN          1     25.0     25.0     50.0
                      GIRL          1     25.0     25.0     75.0
                     WOMAN          1     25.0     25.0    100.0
                              -------  -------  -------
                     TOTAL          4    100.0    100.0
```

A **SORT** command specifying a string can order cases alphabetically. In such a sort, numbers will come before letters and upper-case letters (A – Z) will come before lower case (a – z). Such alphabetic sorting can either be ascending (A – Z, etc.) or descending (Z – A, etc.). When neither an ascending nor a descending sort is specified, the default (ascending) order is used. For example:

```
SPSS/PC: sort by forename.
SPSS/PC: sort by surname.
```

As explained in Chapter 8 (p. 121), when a series of **SORT** commands are issued, special care must be taken with the order in which they are presented. As in the above example, the command that orders the major sort must come *last*. Thus the above pair of commands would lead to a normal 'telephone directory' sort, A – Z by surname, with A – Z sorting by forename *within* groups with a common surname. **SORT** starts with cases in their current order (not in the original order) and does the minimal amount of re-ordering required to satisfy the **SORT** request.

On the **MISSING VALUE** command, the specification for a missing string variable must be specified within apostrophes. Thus:

```
MISSING VALUES var001 to var056 (9) forename "       ".
```

Freefield formatting

In previous chapters, when describing how data should be set out within the data file, specific columns were always allocated to particular variables, and it was stressed that any variable must occupy exactly the same column(s) for all cases.

In doing this, however, we were restricting the data arrangement to the default **fixed format**. There is an alternative format arrangement known as **free format**. In free format, variables are separated by one or more blanks (and/or a comma). Thus the system would automatically recognize the dataset:

```
001 2 34 456 45 5 5 6 4 54 64 23 13 678 78
```

as containing 15 variables. When free format is used, data do not have to be aligned by columns, so that the following dataset would be permissible:

```
001 2 34 456 45 5 5 6 4 54 64 23 13 678 78
002 12 3 46 3 3 2 1 4 46 34 54 12 324 67
003 03 23 324 34 4 5 3 3 41 32 23 9 345 65
004 8 2 23 1 1 2 2 3 47 45 13 23 67 54
```

Each case here contains 15 variables, including *id*. The variables must be in the same order for each case. Although some leading zeros (00 in 001, for example) have been retained in this illustration (and *can* be included in a free format dataset), they serve no useful purpose in a free format situation. Identification numbers, therefore, could be entered as 1, 32, 465, etc. without the leading zeros that are used in fixed format to control the alignment.

If free format is to be used, the keyword **FREE** must be specified after **DATA LIST**. Thus:

```
SPSS/PC: DATA LIST FREE / var001 to var015.
```

The data from more than one case can be entered on a single record, so that the data read as a result of the following command:

```
SPSS/PC: DATA LIST FREE / var001 to var002.
```

might all be contained on the following single record:

```
34 45 12 8 34 56 54 52 45 9 56 56
```

(this would represent the data for six cases with two variables per case).

If the **DATA LIST** command refers to a file, then the command takes the form:

```
SPSS/PC: DATA LIST FILE= "genfree.dat" FREE / var001 to var015.
```

By default, all data are assumed to be numeric, but string variables can be used with a freefield format. String variables must be identified by **(A)** (or the lower case **(a)**) following the variable name. By default, strings are assumed to be eight characters in length. The format can be changed by specifying a width within the **(A)** brackets:

```
SPSS/PC: DATA LIST FREE file="mand.dat"
        :/var001 surname (A15) var002.
```

The corresponding data file ("mand.dat") might contain the following information:

```
321 "Mandeville"  2
7352 "Nolan" 12
 823 "Winterbourne" 23
  45 "Nye" 14
```

the names are shorter than the 15 columns allocated. The system would allocate padding blanks to the right, however, so that the strings would be treated as:

```
"Mandeville     "
"Nolan          "
"Winterbourne   "
"Nye            "
```

If any future reference was made to these strings the user would have to include the extra blanks.

There is little scope for using long strings in any manipulation or analysis, although they can be **LIST**ed and used as **SORT** variables. For example:

```
SPSS/PC: sort by surname.
```

If you attempt to use a long string variable in a procedure such as **FREQUENCIES** or **CROSSTABS** you will receive the warning **'Invalid**

Long String', but the system will carry out the procedure with the variable truncated to eight characters. This generally presents no problem, although when the name 'Winterbottom' was added into the above file, the **FREQUENCIES** output did report the presence of two 'Winterbo' cases in the dataset.

Non-integer data

All of the data used in this book have been in integer form (i.e. there have been no decimal points within the dataset). Sometimes, however (especially with data from experimental studies rather than surveys) some of the data will be decimal data (e.g. 345.566, 23.67, 40.002). There is no problem in including such numbers within a dataset for use by the SPSS/PC+ system, although they will generally represent values for dependent rather than independent variables.

Suppose we have the dataset shown in Table 16 1.

Table 16.1.

var001	var002	var003	var004	var005
23	-345.3	14.004	34	.003
35	45.7	12.234	45	.123
5	142.5	9.030	8	.045
78	-45.6	18.675	27	.067

This set of numbers could be written in a data file (say, "dec.dat") in precisely this form, with the decimal point included. The following **DATA LIST** command might be used to define the variables:

```
SPSS/PC: DATA LIST file = "dec.dat" / var001 8-9 var002 14-19
      :  var003 24-29 var004 34-35  var005 43-46.
```

Fixed format is the default format and need not be specified by the user. The columns allocated to particular variables should make allowance for the decimal point and for the minus sign (-) (each of which would occupy one column).

The above dataset could also be defined by a **DATA LIST** command which specifies a **FREE** format:

```
SPSS/PC: DATA LIST FREE / var001 to var005.
```

If a free format is to be used, the dataset could also be written:

```
23 -345.3 14.004 34 .003
35 45.7 12.234 45 .123
5 142.5 9.030 8 .045
78 -45.6 18.675 27 .067
```

For **FIXED format data lists**, the decimal points may be omitted from the dataset and information added to the **DATA LIST** command to specify how many decimal places should be *assumed* for each variable. The name of a variable for which a decimal point is to be assumed must be followed by the column specification (as usual) and by a number in brackets specifying the decimal places to be assumed to the *right* of the decimal point. This is illustrated in the following example:

```
SPSS/PC: DATA LIST file = "point.dat" / id 1-3 var001 4-5
       : var002 6-7 (1) var003 8-11 (3) var004 12-14 (2)
       : var005 15-16 (2).
```

If the data file "point.dat" contained the following numbers:

```
                  1111111
Columns      1234567890123456
             ================
             9645873443553285
             8568356326443228
             3568932456984675
             2345692345692356
```

it would represent the data for each of the variables in this way:

```
Variables      id  v1  v2   v3    v4  v5
              === ==  ==  ====   === ==
              964  58  73  4435  532  85
              856  83  56  3264  432  28
              356  89  32  4569  846  75
              234  56  92  3456  923  56
```

The system, however, would treat the set of numbers as shown below (following the specifications regarding implied decimal points):

```
Decimal point
specification      (0) (0)   1      3        2      4

  Variables        id   v1   v2     v3       v4     v5
                   ===  ==  ===    ====     ====  =====
                   964  58  7.3    4.435    5.32   .85
                   856  83  5.6    3.264    4.32   .28
                   356  89  3.2    4.569    8.46   .75
                   234  56  9.2    3.456    9.23   .56
```

Decimal data referred to on **FREE format data lists** must include the decimal point. Instructions cannot be given for the system to *assume* a decimal point. The following example shows the contents of a data file ("frepoint.dat") which is to be read following a free format data list command:

```
3.4  23.455  47.567 5.4 2.44
 .76 12.228 23.444 23 44.4
3 78.123   78.993 43 15.4
83 456.66  23.444 78 56.7
43 838.38 56.454 32 523
```

If an additional command, **FORMAT**, is used, the decimal nature of the data can be tidied for presentation in later displays. The **FORMAT** command uses the **F-convention** and is of the form **F4.2**. This example would request that the variable associated with this format specification should be displayed (and written) with four number characters, two of which would be to the *right* of the decimal point. If more decimal places occur in the actual data, then when it is displayed the value of the item is rounded to conform to the specified decimal format. Thus the following commands (referring to the data file "frepoint.dat" above):

```
SPSS/PC: data list free file = "frepoint.dat"/ var001 to var005.
SPSS/PC: formats var001 (F2.0) var002 (f5.1) var003 (f5.3)
       : var004 (f2.0) var005 (f3.1).
```

produced the following output from a **LIST** command:

```
VAR001    VAR002    VAR003 VAR004    VAR005

     3      23.5    47.567      5       2.4
     1      12.2    23.444     23      44.4
     3      78.1    78.993     43      15.4
    83     456.7    23.444     78      56.7
    43     838.4    56.454     32     523.0
```

Data cleaning

It has been assumed that the utmost care has been taken in transferring data from a data sheet, or from the questionnaires, to the computer, and that mistakes have not been made. This assumption may be over-optimistic, however, and the following procedure can be used to detect *some* errors which may have occurred at the data entry stage.

If a variable can take, say, the three valid values 1, 2 and 3, and if that variable has been assigned the missing value 9, then clearly any case which has a value between 4 and 8, or 0, for that variable has been mis-coded, or a mistake has been made in entering the data.

Such errors can be detected by using the **FREQUENCIES** procedure. When you have created a data file and written the relevant data definition file, you can issue the command:

```
SPSS/PC: frequencies all.
```

As the frequency table for each variable is displayed, consult the codebook to determine whether the system is reporting any occurrences of values other than valid and missing values. Make a note of any variables which have such an occurrence.

The next step is to identify which case(s) include an illegal value for any of the variables. Suppose that a variable *triad* has valid values 1, 2 and 3, but that a **FREQUENCIES** analysis has revealed that there is one case with the value 4 for *triad*. Issue the commands:

```
SPSS/PC: process if (triad eq 4).
SPSS/PC: crosstabs id by triad.
```

and you will be presented with a screen display that will identify which case has the value 4 for *triad*. Make a note of this so that after detecting all such errors you can consult the original data source and, having determined which column of the dataset contains the error, you can correct the data file.

No similar procedure will enable you to identify cases for which a value within the valid range has been substituted for the true value, and no cleaning operation such as that described can guarantee that errors do not remain within the dataset. It is very important, therefore, to ensure total accuracy when creating the data file, and to check a printed copy of the file against the original data source.

16.2 Transporting files between computers

Portable files

When a copy of a data file and data definition file, or an SPSS/PC+ system file, has been written to a diskette, it can be used on any other PC computer that has the SPSS/PC+ program. Copies of files can therefore be made for you to use on other machines, or for colleagues to use on systems to which they have access. If a colleague is located at some distance, however, it might be especially convenient to send a copy of a file from your own computer to another machine via a special computer line, or a normal telephone line, using a **file-transfer program**. Files are easily transferred, or transported, between different computers if they are written in a special portable form.

Portable files are created using the command **EXPORT**, and because such files are compatible between SPSS/PC+ and the mainframe computer version of SPSS, SPSS^X, one of the most useful applications of the portable file facility is to create a file from SPSS/PC+ which can be analyzed using SPSS^X, or vice versa. The file-transfer program usually employed to transport SPSS files is named **Kermit**. With **Kermit** in action on both computers and a portable file created on one of them, that machine can send a copy of the file to the other computer. The transferred file can then be read using an **IMPORT** command. Portable files are not in the standard text file format and, generally, should not be edited. When an **IMPORTED** file has been made active it can be **SAVED** as a system file.

The EXPORT command

EXPORT is used to create a portable file from the current active file. The portable file is written to the current directory and can later be transported (for example, using the file-transfer program **Kermit**) from SPSS/PC+ to another system. If a portable file is to be uploaded to another computer, suitable (and compatible) file-transfer programs must be loaded on *both* machines. **EXPORT** can also be used on another machine (for example, a mainframe computer running **SPSS^X**) to create a portable file to be exported (via **Kermit**) to the computer you are using.

EXPORT writes the current active file, including the data itself, the data dictionary, and certain variables assigned by the system. Although portable files are written in alphanumeric (ASCII) format they are not easily editable.

The simplest **EXPORT** command includes the command word followed by **outfile=** and the name of the portable file to be created. Thus:

```
EXPORT OUTFILE= "gender.por".
```

In this case the entire contents of the active file would be placed in a new export file named "gender.por". The variables would retain their original names and values together with the original variable labels, value labels and missing value flags. Changes to the active file made as a result of any **COMPUTE**, **RECODE**, **SELECT** and other commands issued during the session will remain in effect when a portable file is created.

It is possible to write a portable file that includes only *some* of the variables in the active file. This can be done either by specifying the variables to be retained (using a **Keep** subcommand) or those to be dropped (using a **Drop** subcommand). Thus:

```
SPSS/PC: EXPORT OUTFILE= "gender.por" / KEEP= id to marit.
```

or

```
SPSS/PC: EXPORT OUTFILE= "gender.por" / DROP= age mar01 to e12.
```

The portable file created by the second of these examples would include all of the variables on the active file *except* the variables listed.

Notice that the **TO** convention is permitted in **EXPORT** (and also **IMPORT**) commands. You may remember that it was *not* permitted within **JOIN** commands.

Judicious use of a **Drop** or a **Keep** command may be essential when creating a portable file on a mainframe for later transportation to SPSS/PC+. A mainframe has far more capacity and power than a microcomputer, and if the portable files produced from $SPSS^X$ are too large then the microcomputer may not have sufficient memory capacity to deal with them. A portable file created by SPSS/PC+, or a portable file to be read by SPSS/PC+, cannot contain more than 500 variables.

Variables in the active file may be renamed (using the subcommand **Rename**) as the portable file is created. Thus:

```
SPSS/PC: EXPORT OUTFILE= "gender.por"
       : / rename= (st01 st03 = depress anxious).
```

Here *st01* would be renamed *depress*, and *st03* would be renamed *anxious*.

If changes are made to the file (with **Drop**, **Keep** or **Rename**) it is useful to keep within the portable file a listing of the changes. This will help the user to keep track of how the portable file relates to the original file from which it was created. Such a listing is requested by including **Map** as a final subcommand:

```
SPSS/PC: EXPORT OUTFILE= "gender.por"
       : / rename= (st01 st03 = depress anxious)
       : / drop= dat01 to dat08 / map.
```

Changes made in creating a portable file do not affect the active file.

Having created an exportable file the user will probably want to send this file to another computer using **Kermit**. Before considering how this is done, however, we will consider the use of the **IMPORT** command. This is used to read a portable file into the SPSS/PC+ system and is similar to **EXPORT** in many ways.

The IMPORT command

IMPORT is a 'data read' command (i.e. a data definition command rather than a procedure) which reads information from a portable file. When a subsequent procedure command is issued the active file is created. The portable file must be in the current directory on the hard disk, although it can be imported from a floppy diskette in either of the diskette drives, *Drive A* or *Drive B*. Typically the portable file will have been downloaded, via **Kermit**, from a mainframe computer running $SPSS^X$.

The simplest **IMPORT** command includes the command word followed by **file**= and the name of the portable file to be read. Thus:

```
SPSS/PC:  IMPORT FILE= "mainfram.por".
```

In this case the entire contents of the portable file would be read. To read only some of the variables within the portable file a **Drop** subcommand or a **Keep** subcommand would be used, specifying those variables to be dropped or kept. Thus:

```
SPSS/PC:  IMPORT FILE= "mainfram.por" / DROP= item45 to item54.
```

If **Drop** is used then the variables will appear in the active file in the same order as they occur in the portable file. If **Keep** is used, however, the order in which the variables are named will determine their order within the active file.

Variables may also be renamed as they are read from the portable file, using a **Rename** subcommand. Thus:

```
SPSS/PC: IMPORT FILE= "mainfram.por"
       : / rename= (item1 to item44 = var001 to var044).
```

This would simply rename all 44 of the *itemX* variables as the equivalent *varXXX* variables. Thus *item1* would become *var001*, *item41* would become *var041*, etc. The above command also illustrates the fact that the **TO** convention is permitted within **IMPORT** commands.

As with **EXPORT**, a listing of any changes made during the **IMPORT** (with **Drop**, **Keep**, or **Rename**) can be included in the file. Once again, this is done using **Map** as the final subcommand:

```
SPSS/PC: IMPORT FILE= "mainfram.por" / rename= (item45 = total)
       : / drop=item20 to item23 / map.
```

Having discussed how an exportable file can be created with **EXPORT** and how an importable file can be read with **IMPORT** we now need to consider how such files may actually be transferred from one computer to another.

Using KERMIT to transfer portable files

Two computers may be linked by a special line, or the two computers may be linked via a telephone link, each using a piece of apparatus known as a **modem**. The 'host' computer is dialled from the resident or 'home' machine. Using the appropriate commands the user introduces him/herself to the host machine (using log-on instructions, special passwords and the like) and information can then be transferred between the two machines. Commands issued from the keyboard of the resident computer can be made to operate programs in the remote machine.

The file-transfer program **Kermit** is generally used to transfer portable SPSS files. A file created by **SPSS**^{X} on a mainframe computer can be transported to a SPSS/PC+ microcomputer and vice-versa. **Kermit** is error correcting, so that after a successful file transfer the user can be sure that the copy is a true replica of the original.

To access the mainframe computer you will, of course, need to be registered as a user. Certain parameters of the communication line will need to be appropriately set (these include the baud rate – the signalling speed – and the parity – an error-checking strategy). Because the setting of certain parameters is crucial to the success of file transfer, your first attempt at using **Kermit** may not prove entirely straightforward. If other users working with your system have experience in the successful transportation of files to and from the other computer then they may be willing to demonstrate the process to you and to explain how the particular link functions. Many different versions of **Kermit** are available and they are variously implemented on most mainframe computers. Compatible versions of **Kermit** need to be available on the mainframe and the microcomputer.

To operate file transfers from the SPSS/PC+ microcomputer, the **Kermit** program is run (through **DOS**), and a 'connect' command issued. The appropriate telephone number is dialled to connect to the mainframe computer and, after an acknowledgement from the host computer has been received, the user logs-on to the host.

At this stage the resident machine acts as a terminal of the mainframe. The host computer can be requested to run its own **Kermit** program to send a portable file, for example:

```
SEND mainfram.por
```

The user then escapes back to the microcomputer (for example, by pressing **<Ctrl-]>** and then **<C>** – a screen display will usually provide information on the appropriate escape sequence).

The microcomputer **Kermit** program is then instructed to receive a named portable file which has previously been created on the mainframe by using the **EXPORT** procedure. For example:

```
RECEIVE mainfram.por
```

Transfer should now occur, and the requested portable file should be written to the current directory of the microcomputer. The user now regains connection with the host computer in order to log-off and end the connection to the mainframe.

When the portable file is **IMPORTED**, using a command such as:

```
SPSS/PC: import file= "mainfram.por".
```

it becomes the active microcomputer file, ready for transformation and analysis. And once the information from the portable file has been made active it can be **SAVED** as a system file.

The process of uploading a portable file from a microcomputer to a mainframe is very similar, except that the **SEND** and **RECEIVE** commands are issued to the microcomputer and the mainframe respectively. After transfer, a copy of the portable file should now be in the target location within the mainframe computer memory. Hopefully this process will prove successful even on the first occasion, but if there are problems you should seek advice from your computer advisory service.

16.3 Expanding your SPSS/PC+ repertoire

This is an introductory book. It is designed to enable first-time users of the SPSS system to become acquainted with the fundamentals of the microcomputer-based SPSS/PC+ package. No attempt has been made to present a comprehensive account of even the base system. Many of the more complex aspects have been presented in a simplified form or avoided altogether (especially if they are seldom used) and a number of procedures have been completely omitted. Furthermore, no attempt has been made to cover the many further procedures available within the various SPSS/PC+ modules – *Advanced Statistics*, *Graphics*, *Trends*, etc. As your familiarity with SPSS/PC+ develops you will certainly wish to explore some of these.

The final part of this chapter provides an indication of some aspects of SPSS/PC+ that you might wish to add to your repertoire with the aid of the *SPSS/PC+ 4.0 Base Manual* and the manuals written for the various optional modules. The intention is to offer just a brief description so that you can decide for yourself which additional facilities are likely to be of the most interest and benefit to you.

The base system

Fifty different commands are available within the SPSS/PC+ base system, and over half of these have been covered in earlier chapters. We have examined the central core of the base system; some of the commands we have not discussed relate to special functions which only a minority of users would wish to take advantage of (such as reading files from 'foreign' database programs, or linking with specific graphics programs). We have ignored other commands because they do much the same thing as commands that we have examined, or because they do things which might be easily accomplished in some other way (for example, by using a word-processor). Finally, some commands have been omitted from detailed consideration simply because the inclusion of too many commands would be likely to confuse those who are coming to SPSS/PC+ for the first time.

You may by now be so impressed with the system, however, that you wonder what else can be done. In this section I provide a brief description of each of the commands (in the base system) not otherwise covered in this book. After reading these brief descriptions, presented in alphabetical order of the command, you will therefore have some idea of the function of every command available in the SPSS/PC+ 4.0 Base system. If you wish to use any of the following commands, however, you will need to consult the *SPSS/PC+ 4.0 Base Manual*.

ADD VALUE LABELS: a transformation command which adds or changes value labels for one or more variables without deleting all existing value labels for the variable(s) (as issuing a new **VALUE LABELS** command would). Unlike most transformation commands, **ADD VALUE LABELS** takes immediate effect (i.e. the data do not have to be read first).

AGGREGATE: a procedure command which creates a new active file or system file by aggregating groups of cases into a single summary case. Thus:

```
SPSS/PC: AGGREGATE outfile = "sexgroup.sys" / break = sex
       : / sexg006 = mean(var006).
```

would create a system file named "sexgroup.sys", which would contain just two cases (male and female; this follows from the **Break** subcommand which requests that cases be aggregated in terms of their value on the variable *sex*). The variables included in the file would be *sex* (one of the cases would probably have the value 1, and the other case the value 2) and a new variable *sexg006* which is an average of all *var006* values for each of the two original groups. However, the above command would only produce a system file of the type described if the active file contained all male cases together and all female cases together. Otherwise the command would form groups by adjacent cases with the same value for the break variable. A **Presort** subcommand (specified *before* the **Break** subcommand) will sort cases in terms of the **Break** variable if they are not already in this state. Thus:

```
SPSS/PC: AGGREGATE outfile = "sexgroup.sys" / presort /
       : break = sex / sexg006 = mean(var006).
```

Many different aggregate functions are available for the creation of new variables from the original variable values. 'Mean' was used in the example; others include 'sum' and various percentage functions.

AUTORECODE: a transformation command which creates a new variable from an original variable, recoding the original values into consecutive integer values. Thus if cases in the dataset had values, for an original variable *option*, of 13, 16 and 18 (only), the automatically recoded variable (it might be named *roption*) would have 1 as the value for cases which have the *option* value 13, 2 for cases which have the *option* value 16, and 3 for the *option* value 18. The command might be:

```
SPSS/PC: AUTORECODE variables = option
       : / into roption.
```

Lists of variables can be included in the command (the **Into** subcommand must provided a name for each new variable). AUTORECODE is particularly useful for getting rid of empty cells in crosstabulations, and for recoding string variables (integer values are provided in place of the strings which are, in effect, sorted alphabetically). The assignment of integer values may be by ascending order of the original (numerical or alphabetic) values (the default), or by descending order.

BEGIN DATA/ END DATA: commands used immediately before and immediately after data which are included within a data definition file. However, such an 'in-line' arrangement is not recommended. When a separate data file is used, as shown in this book, the commands are *not* used.

* (COMMENT): a facility that allows you to insert commentary lines either when you are editing a command file or interactively at the SPSS/PC: prompt. Provided that the comment does not occur within data lines or within a command, the operation of the system is not affected by such lines, and the comments appear in the log (and listing) file. To specify that a line is a comment, it must begin with an asterisk in the first column, followed by a space. Text may follow on this line and succeeding lines (which do *not* need an asterisk). The final line of a comment must end with a period, but no other line of the same comment may end with a period.

COUNT: a transformation command which creates a new variable for each case by counting across several variables. Thus if 20 variables (Q1 to Q20) were coded 1 for 'yes', 2 for 'don't know' and 3 for 'no', a new variable (*dontknow*) could be created to record the number of times the subject ticked 'Don't know' boxes:

```
SPSS/PC: COUNT dontknow = q1 to q20 (2).
```

DESCRIPTIVES computes descriptive statistics for variables. Most of the facilities it provides are also available with the **FREQUENCIES** procedure (and the **MEANS** procedure), although it also has a useful **OPTION** for adding standardized scores (i.e. *Z* scores) to the active file.

EXAMINE: a procedure command used to provide a range of descriptive statistics, to provide graphic output in the form of histograms, stem-and-leaf plots and boxplots, and to test whether the data are normally distributed.

EXECUTE: a command which allows the user to run other programs and to issue **DOS** commands within the SPSS/PC+ session. However, limited memory is available for the operation of other programs and some programs may cause SPSS/PC+ to abort. In most cases it is preferable to run other programs after ending an SPSS/PC+ session. In earlier chapters, whenever **DOS** commands were used to copy or print a file, for example, this was always done before or after an SPSS/PC+ session.

FLIP: a command which transposes rows and columns, so that the columns become the rows and the rows become the columns. SPSS/PC+ works with files in which the rows are the cases and the columns are the variables. If a dataset with the alternative arrangement is to be used by SPSS/PC+, **FLIP** must first be used to create a reorganized file.

FORMATS: a transformation command which specifies how numeric values are printed and how information is formatted when a **WRITE** command is issued. The values used in computations are unaffected by **FORMATS** specifications.

GSET: a command used when working with one of a number of widely used graphics packages including *Harvard Graphics*, *Microsoft Chart* or *Graftalk*.

MODIFY VARS: a command used to change the contents of the current active file by re-ordering variables, dropping some variables or specifying which are to be retained in the file, or renaming specified variables. The subcommands **Reorder**, **Drop**, **Keep** and **Rename** are used, and **Map** as a final subcommand will produce a table showing the names and order of variables within the active file.

REPORT: a command used to list values for individual cases, or to generate tables of summary statistics for groups of cases. The command provides a flexible means of tailoring the presentation of the table (in terms of labelling, titling, column width, and spaces between lines). Many subcommands are available (a number are *required*) and the order of the subcommands is important. The procedure also assumes that cases within the dataset are arranged in a suitable sequence, so a **SORT** command usually needs to be issued before **REPORT**. Using **REPORT** is not easy, and the command section information on this procedure in the *SPSS/PC+ 4.0 Base Manual* is considerably longer than that for any other command. Many users find it preferable to produce tables (for inclusion in reports, etc.) by word-processing the listing output from other commands (including **MEANS**).

SPSS MANAGER: a command that allows the user to install or remove groups of SPSS/PC+ procedures to or from the hard disk. SPSS/PC+ procedures (both those from the base system and those from the option packages) are grouped together in modules, and the unit of installation and removal is the whole module. A little used module can be removed to create more space on the hard disk, and then installed from the original diskette copy (or a backup copy) when a procedure within the module is needed.

TITLE and **SUBTITLE**: these commands are used to specify text to be written at the top of each page of output. Thus:

```
SPSS/PC: TITLE "First Gender Survey Analysis Session".
SPSS/PC: SUBTITLE "Frequencies - MALE respondents only".
```

The title text provided by the user can be up to 58 characters long and is written in the centre of the top line of each output page until the end of the session or until another **TITLE** command is issued. Title text does not need to be enclosed within apostrophes or quotation marks, but if it is not then it is converted to upper-case. The same rules hold for the **SUBTITLE** command. Subtitles can be up to 64 characters long and are written at the left-hand side of the second line of each page. The commands are independent, so that one can be changed without affecting the other.

SYSFILE INFO: a command which displays information about a specified SPSS/PC+ system file, including the file name and the date and time when it was created, the number of cases, variable names and labels, and information about system variables and certain system characteristics of the file.

TRANSLATE: a command which creates an SPSS/PC+ active file from (or converts an SPSS/PC+ active file to) a 'foreign' file from spreadsheet and database programs, including several of the most widely used. **From** and **To** are subcommands. Thus to create an active file by translating from a *dBase IV* file named 'gen.dbf', the following command might be used:

```
SPSS/PC: TRANSLATE From = "gen.dbf"/ Type=db4.
```

A **Drop** or **Keep** subcommand can be used to include only some of the variables from the original foreign file in the SPSS/PC+ active file. **Map** can be used to produce a list of variables with some of their characteristics. If a file read with **TRANSLATE** has the variables as rows and the cases as columns, **FLIP** procedure should be used immediately after **TRANSLATE** to reverse this situation.

WEIGHT: a transformation command which affects the system variable *$weight*. The initial value of this is 1.00 for each case. The command tells the system to assign different weightings to cases. Thus the command:

```
SPSS/PC: WEIGHT by sex.
```

would, if males were coded 1 and females 2, replicate each female case twice in subsequent procedures, thus inflating the size of the sample (the effect of such a change in weighting could be observed by issuing a subsequent **FREQUENCIES** command). The *$weight* variable may be re-set to 1.00 during a session by issuing the command 'WEIGHT off'. Otherwise *$weight* remains with the assigned value and will be **SAVED** within a system file or within an **EXPORTED** portable file.

WRITE: a procedure command which writes cases within the active file to an ASCII results file "SPSS.PRC" (this default name may be changed by a **SET** command) and displays a table showing the names and location of variables within the file. Records (lines) are 80 characters in length and are formatted with a single blank space between variables. **WRITE** can create a file of all or just some of the variables and cases from the active file. System variables are included only if an explicit request is made.

SPSS/PC+ Graph-in-a-Box
This is an extra facility which comes with the base system. It is a memory resident program which captures data from a screen display of part of the listing file (taking a 'snapshot') and converts the numbers into a barchart, a line chart or a pie chart.

The optional modules

The base system includes commands for handling data and files, and includes a number of statistical and report procedures. It is needed to run any of the many modules that are now available and are distributed separately. One module, *SPSS/PC+ Statistics* is indispensable, and several of the procedures described in this book are included within that package. The other modules may also be necessary to your requirements, depending on whether, for example, you need to work with complex multivariate statistical methods and whether you need to produce high quality tables and graphs.

SPSS/PC+ Statistics
This includes 10 procedures, six of which have been described in this book.
We have met:

T-TEST: Chapter 7.
CORRELATION: Chapters 11 and 13.
ONEWAY: Chapter 12.
ANOVA: Chapter 12.
NPAR TESTS: Chapter 13
RELIABILITY: Chapter 14

We have not discussed **REGRESSION**, **FACTOR**, **CLUSTER** and **QUICK CLUSTER**.

REGRESSION performs linear and multiple regression analyses. Linear regression is a relationship between two variables when their values, plotted graphically, produce a straight line. If a straight line is not a good representation of the relationship, linearity may be achieved by transforming data. Values for the dependent variable can be predicted from scores on the independent variable by an equation which represents the slope of the line and the point at which it intercepts an axis on the graph. Multiple regression is a technique for estimating simultaneous correlations between multiple predictor variables and a single dependent ('criterion') variable. It therefore allows the prediction of a value for the criterion variable, for a particular case, from the values of the predictor variables. The **REGRESSION** procedure calculates the relevant equations and can produce a variety of graphical plots. A number of alternative regression methods are available. The procedure can be used with matrix materials (for example, with a results file obtained from the **CORRELATION** procedure) as the input.

FACTOR performs factor analysis, a statistical technique which computes, from a correlation matrix, the minimum number of 'factors' which will account for the overall pattern of inter-correlations. The process is somewhat similar to multiple regression, except that variable values are predicted from the values of hypothetical 'factors'. Factor analysis is used, for example, to identify the basic factors being measured by a large number of items within a personality inventory. Whereas the item responses are directly observable, the underlying factors which might have produced the pattern of results are not. The analysis helps to identify these underlying factors. Factors are first extracted and then transformed (by a mathematical process labelled 'rotation') until they are easily interpretable, and the ideal endpoint is a 'factor solution' which can represent a complex correlation matrix in a simple, easily interpreted, form. If this is the situation, scores on the key factors can be computed for each case, so that factor analysis provides the basis for scoring the instrument. **FACTOR** provides a choice of many different methods for extracting and rotating factors, and allows factor scores to be computed for each case.

CLUSTER performs cluster analysis, a type of multivariate analysis which identifies groups (clusters) of similar cases. Similarity or closeness is judged on the basis of the values of cases for a set of variables. After initial clusters have been identified, **CLUSTER** combines clusters and then computes new distances. The process is repeated until all cases form a single cluster. The result is a hierarchical analysis, and the graphical output may include a tree-like structure (a dendrogram) linking cases. Many different methods of measuring distance, and many different clustering methods, are available within the procedure. The cluster membership of cases can be saved as a new variable in the active file, and a matrix of the distance between cases or of the similarity between cases can be written to a results file.

QUICK CLUSTER is a less flexible procedure, but is faster and needs less memory. The user specifies how many clusters are to be formed. Initial cluster centres are identified either from the first cases or from cases that have substantially different values, and these centres are then moved (updated) as each new case is examined. After the final cluster centres have been determined, each case is examined again and assigned to the most appropriate cluster. The results of the analysis can be saved in the form of two new variables for each case, one identifying the cluster to which the case belongs, and the other specifying the distance of the case from the centre of its cluster. Cluster centre data can also be written to disk as a results file.

SPSS/PC+ Advanced statistics

This includes eight procedures, none of which have been described in this book.

DSCRIMINANT performs discriminant analysis, a technique for predicting group membership on the basis of values on a set of predictor variables. The procedure involves classifying cases into groups by using the discriminant functions that best distinguishes between the groups. (Note: **DSCRIMINANT** is 'mis-spelled' so that the command can be distinguished from **DISPLAY** when the command is issued in 'truncated' form (i.e. just the first three letters).

MANOVA performs multivariate analyses of variance, including those which apply to univariate research designs (i.e. with just one dependent variable), multivariate designs (with two or more related dependent variables) or repeated-measures designs. The last of these designs involves the same variable being measured for the same case on several different occasions. It has the advantage that error variance due to differences between subjects is eliminated.

LOGLINEAR performs a number of related techniques for the analysis of multidimensional crosstabulations. The variables which specify the dimensions of the crosstabulation are the independent variables, and the cell frequencies together comprise the dependent variable.

HILOGLINEAR analyses multidimensional crosstabulations, fitting hierarchical models to the data.

LOGISTIC REGRESSION performs a regression analysis with a *dichotomous* dependent variable and a set of independent categorical or numerical variables.

NLR (non-linear regression) performs regression analyses using complex non-linear models. The procedure estimates goodness-of-fit statistics and can compute predicted values.

PROBIT performs probit and logit analyses. These estimate the effects of independent variables on a dichotomous dependent variable. Thus the effects of exposure to different amounts of T.V., leaflet, and newspaper advertising could be estimated for the single dichotomous variable reflecting whether the individual had bought or not bought the product. **PROBIT** tests a prediction model.

SURVIVAL produces tables and graphical plots that describe the temporal relationship between two events (such as marriage and divorce, medication and cure, birth and death). Data is entered as a summary of survival in a population as assessed at regular intervals, and from this the procedure computes a 'life table'. This begins at the first point in time with the original population size, and then displays statistics about the population as its size diminishes over regular time intervals. The table can be used to estimate the probability of a case leaving the group of survivors at any point in time, thus providing a guide to the likely fate of those still surviving in the population.

SPSS/PC+ Trends
This module includes a comprehensive set of procedures for forecasting and time-series analysis. It allows the user to plot changes over time, to derive models of change, and to predict future trends.

SPSS/PC+ Categories
This module provides procedures for the complex analysis of categorical data. It particularly useful for analysing, for example, consumer perceptions and preferences regarding different products. The types of analysis available include conjoint analysis and correspondence analysis.

SPSS/PC+ Tables
This module allows the user to produce high quality summary tables which can easily be incorporated in final reports, or used as the source for making slides and overhead projection materials. It gives the user great freedom in determining the format and content of tables. Titles and footnotes can occupy several lines. The module also allows several tables to be combined within a single display.

SPSS/PC+ Graphics
This module enables the user to produce high quality graphics to display features of the data. For example, the **PIE CHARTS** procedure will enable the frequencies for various values of a variable to be displayed as suitably sized segments of a circle, each appropriately labelled and coloured (or shaded with a different design). Pies can be rotated to any angle and the segments can be arranged in any order. A similar degree of flexibility is possible with bar charts, line graphs and scatterplots. The user is able to switch instantly between alternative ways of presenting the data and thus to choose which type of display presents the best portrayal of the relevant features of the dataset. *SPSS/PC+ Graphics* features *Harvard Graphics*, a best-selling graphics program which includes a library of over 300 images (symbols, logos, borders and line drawings) that can be used to customize the statistical charts.

SPSS/PC+ Map
This module enables the user to produce displays of geographical maps which can be differentially coloured, etc. to reveal, for example, the

distribution of a particular population characteristic.

SPSS Data Entry II
This provides an alternative way of entering data into a dataset. Data entry forms are created, and the module is used to clean old or new data, to modify data and to translate files from one format to another. *SPSS Data Entry II* is particularly useful when there are a large number of cases in the dataset and when a questionnaire contains 'branches' (when questions do not apply to a particular respondent a skip-and-fill facility is used). However, setting up the initial template is something which beginners may find rather complex, and the use of this module is not recommended in the initial stages of learning to use SPSS/PC+, particularly if the first datasets to be analyzed are of modest size.

SPSS/PC+ CHAID
This is a stand-alone package that performs Chi-square Automatic Interaction Detection. This enables the user to explore data before running other procedures, so that extraneous variables may be excluded, and the number of categories for categorical variables may be reduced. *SPSS/PC+ CHAID* can then be used to produce predictor diagrams and decision diagrams based on various predictor models. The module is of special relevance to those in the consumer field who need to predict consumer choice and to analyze the effects of pricing and packaging.

DBMS/COPY Plus
This module translates between SPSS/PC+ system files and over 60 popular database, spreadsheet and statistical software packages. Files can be translated from the foreign package to SPSS/PC+ and vice versa, and selections of variables and cases can be made. Files translated with *DBMS/COPY Plus* retain their necessary dictionary information.

SPSS/PC+ Fast Start
This is a computer-based training module which includes teaching on a few simple procedures and on the use of the menu and help systems. There is also guidance on the use of **DOS** commands.

SPSS/PC+ Codebook
This is an aid to producing a user's codebook. It allows the printed codebook to be formatted to the users specification, defining the variables and the codes and documenting the coding instructions for each question.

16.4 A final word

The purpose of this book has been to introduce SPSS/PC+ in the style of a tutorial rather than to provide a comprehensive manual. Short cuts have been taken, and many statistically advanced topics have been avoided. Having covered the material available in this introduction, however, you should be able to follow the detailed presentation of all of the topics covered in the *SPSS/PC+ 4.0 Base Manual* and in the manuals provided with each of the various modules.

It is certain that in the next few years the SPSS/PC+ system will develop further, and that new modules will become available. The core of the SPSS/PC+ package, however, is unlikely to change substantially, and the skills and concepts you have learned in working through this book should therefore remain useful to you for a long way into the future.

APPENDICES

Appendix A: Notes for Trainers

Appendix B: Using Word-processor and Other Files with SPSS/PC+

Appendix C: The Gender Attitudes Survey: QUESTIONNAIRE

Appendix D: The Gender Attitudes Survey: CODEBOOK

Appendix E: The Gender Attitudes Survey: DATA

Appendix F: The Gender Attitudes Survey: DATA DEFINITION FILE

Appendix G: Glossary of Statistical, Computing and SPSS/PC+ Terms

Appendix H: Types of File Used or Created by SPSS/PC+

Appendix I: Variations between Different Versions of SPSS/PC+

Appendix J: Brief guide to SPSS/PC+ Commands and Subcommands

Appendix K: Brief Guide to REVIEW Commands

Appendix A: Notes for Trainers

The method presented in this book is one of many possible ways for teaching beginners to use SPSS/PC+. The particular approach has developed after many years' experience in teaching undergraduate and postgraduate students how to use various versions of the SPSS system. SPSS/PC+ itself has become something of a 'moving target'. As it has evolved through successive versions the system may have improved as far as regular and experienced users are concerned, but some of the changes have done little to make the system easier for beginners to learn.

This book offers a 'narrative' presentation. Readers are led through the process of designing a suitable questionnaire, and producing a full codebook specification, *before* they begin to create a data file and a data definition file. They are then taught how to issue commands to the system to analyze and transform data.

The method presented is not particularly sophisticated or elegant, and it may not draw upon the full potential of SPSS/PC, or maximise processing speed. However, for the novice these are minor concerns. SPSS/PC is a tool, not a musical instrument, and students are interested in analyzing their data rather than developing virtuosity with the system. Of course they should be aware of the full potential of the system, and of course they should be encouraged to augment their skills in using the system. In the final chapter, therefore, I introduce a number of alternative ways of working, and briefly describe the full range of SPSS/PC+ facilities. But it is important that students should not be intimidated by the power and potential complexity of the system and that the initial steps are taken slowly and simply.

My aim in this book is therefore to get the user 'up and running' with the minimum possible learning time. This is achieved by keeping the presentation of procedures plain and simple, and by describing the operation of the system so that it remains 'transparent' to the user.

It is instructive to compare learning SPSS/PC with learning to type. If someone intends to be a 'copy typist' capable of working at the greatest possible speeds, then they should start by learning the 'correct way', because learning to type with just two fingers will interfere with the development of proper 'touch typing' skills. Learning SPSS/PC is quite different in this respect. Learning the easy way will provide the user with a good understanding of the system and will not interfere if, at a more advanced stage, they wish to realize the full potential power and speed of SPSS/PC+. And, like typing, the 'finished product' (the letter, or the statistical analysis) will be essentially the same whether it is a product of simple or expert technique.

Another major distinction between learning to type and learning SPSS/PC+ is that typing skills are likely to be used every working day whereas SPSS/PC+ will probably be used only sporadically. Continual use means that skills are constantly rehearsed and become automatic, but few people use SPSS/PC+ every working day. Most take some months to complete a research project, and then use SPSS/PC+ intensively for a short while. There may be intervals of weeks or months between periods of concentrated SPSS/PC+ use, and during these intervals even the very experienced user may forget some basic elements of the system. Re-learning will be more rapid if the user has a fundamental and well-structured understanding of a simplified system. The clear and simple methods used in this book therefore make it suitable as a brief revision text, and I have found that even expert SPSS/PC+ users find both the text and the appendices useful aids to revision and re-learning.

This book is 'dedicated' to the PC microcomputer version of SPSS. Unlike other books which introduce SPSS/PC+ I have not attempted to write at the same time for those who are using SPSS^X. Neither have I tried to cater for those who are working with the newly released Apple Macintosh version of SPSS. In my experience, texts which attempt to simultaneously address two or more versions of SPSS confuse many of those struggling to understand their own version of the system for the first time.

The method used in this book

The method for using SPSS/PC+ presented in this book involves bypassing the initial menu (using **<Alt-F10>**) and then working from the SPSS/PC: prompt. This is the 'home base' from which commands are issued. In response to the prompt the user can enter REVIEW, or INCLUDE a file, or issue a procedural or operational command. When one operation has been completed, the user 'returns to base'. Thus although it is not necessary to return to the prompt when a command file has been edited, for example (commands can be run directly from within REVIEW), this 'makes sense'. In such a situation, the student is instructed to edit a command file with REVIEW, to save it, to exit from REVIEW, and then to INCLUDE the command file in response to the SPSS/PC: prompt. Students therefore easily remember where they are in a system that can all too easily take on some of the characteristics of a labyrinth.

Data are entered into a separate data file (using REVIEW – *SPSS/PC+ Data Entry II* is *not* used) and are not included within data definition or other files. Furthermore, all data are coded in numerical form and written in fixed format data files that contain only integer data. Discussion of free format, and the use of strings and non-integer data, is left until Chapter 16. A separate data definition file is used to define the data.

Initially, all commands are issued interactively from the SPSS/PC: prompt. I encourage readers to work with files that remain editable, rather than with system files, and I emphasize the use of the listing file (SPSS.LIS) in collecting the results of procedures at the end of a session.

Although no attempt is made to examine in any detail any of the powerful multivariate analyses possible with SPSS/PC, these are briefly described in the final chapter. It is my belief that when the reader has mastered the procedures included in this book, he or she will have developed a thorough understanding of the basic system and will be in a good position to follow the detailed information provided in the *SPSS/PC 4.0 Base Manual*, the *SPSS/PC+ Statistics 4.0* manual, the *SPSS/PC+ Advanced Statistics 4.0* manual, and the manuals provided for the other optional modules.

The rationale

In this book I make little reference to the extensive menu system and the help system because experience has taught me that many beginners often find these 'aids' confusing and intimidating. The on-line help facility may be helpful for the experienced user, but its use demands some familiarity with complex menus and menu commands. The menu structure is six levels deep, and navigating through them can be as engrossing(!) as exploring a computer adventure game. Anyone who uses SPSS/PC+ every day might find it worthwhile to build a cognitive map of the labyrinth, but this is unlikely to be the case for the majority of students and research workers are only 'occasional users' of SPSS/PC+.

Throughout most of the book, I avoid the use of systems files. The use of such files increases speed and offers new possibilities (for joining files, for example), but for the beginning student working with relatively small datasets these benefits are outweighed by the costs. By working with editable files, rather than system files, the student will remain 'close' to the data and the data definition information, and will retain a sense of control over the file contents. The increased speed offered by working with system files may be of little importance to students, and the typical PC in use today is substantially faster than those available when SPSS/PC+ was first introduced in the early 1980s.

How you can help the student

What can you do, as a trainer, to help students who are using this book as their main introduction to SPSS/PC+? There are two main ways in which you can help. One is to familiarize yourself with the method and terms presented in this book, so that you 'speak the same language'. If you are advising students to include data within the data definition file, or to work exclusively with system files, while this book encourages the use of separate files that remain editable and are directly printable, then the student is likely to become confused. Even if the method you use personally is somewhat different, remember that the novice needs a more elementary approach and, above all, needs information to be coherent and consistent.

The other way in which you can help is by making the 'front end' of the system 'user friendly' (in this case this means compatible with the instructions given in this book). In Chapter 3 it is suggested that each user should create a personal directory on the hard disk and work from that directory. When examples of DOS commands are given in the text, they are shown as being issued from the C> prompt. Ideally, there would be enough space on *Drive C* for this to be possible. If you require that students create a personal directory on *Drive D* or above, or if you wish them to keep all of their personal files only on a floppy disk, you will need to provide clear instructions about how their procedures should differ from those suggested in this book.

Many students are technophobic and need to be given considerable encouragement before they approach the computer. It would be a great pity if they were to meet problems and encounter complications before they even get into the SPSS/PC+ system. You can help them to avoid confusion by making entry to the system straightforward. Ideally, there would be sufficient space on *Drive C* for many personal directories (whether or not the SPSS/PC+ program itself is written on that drive), and a path command would be included in the autoexec.bat file so that SPSS/PC+ can be used from any such personal directory. Ideally, each student would be provided with a preformatted floppy disk onto which they could copy their personal files at the end of a session.

There are many differences in computer keyboard designs and the labelling of particular keys, and this can be a further source of confusion for those who are unfamiliar with computers. It may be worth checking to see whether the keyboards of the machines your students will be using differ in any important respect from those described in the text. Similarly, depending on which keyboard program is in operation, the symbol that appears on the screen when a particular key is pressed may not be that which would be expected from the label on the key. Thus apostrophes (") may be produced by pressing the **<@>** key. Unless students have been told of such a transposition, they may spend some considerable time hunting on the keyboard for a key which will produce the required symbol on the screen. Either by loading a different keyboard program or physically pasting a different label on the key this potential source of frustration can be eliminated.

On your personal machine, it is useful to tailor the system to your individual requirements, for example by writing a complex 'SPSSPROF.INI' file with a SET command that changes operational features of the system, including the default names for output files, the on/off status of MORE and PRINTER, the WIDTH, etc. However, on a machine in general use within a college department or agency, such changes can confuse the novice, because the system will not then do what, according to this book, it *should* do. This book assumes that all of the operating characteristics of the system remain at their initial default setting, and it would therefore be preferable to keep such setting on the machine(s) you use for teaching.

Inevitably, however, the limitations, machine characteristics and configurations of some teaching installations will require some departure from the 'ideal' system set-up suggested here ('ideal' here means 'ideal for beginners using this book to become conversant with SPSS/PC+'). If this is the case in your teaching environment, you can help by setting the parameters as close as possible to those indicated in this book and produce a simple step-by-step guide giving clear instructions of the action to be taken to gain access to SPSS/PC+, to create a directory or work from a floppy disk drive, or to copy files. You should also make it clear that your directions override particular instructions presented at the beginning of this book. A printed copy of the new set of directions might be given to each student, or a notice might be displayed prominently near the machine(s) that students use.

Note to trainers familiar with the first edition

This second edition of the book has provided me with the opportunity to rewrite sections on certain procedures that continue to give beginners problems, as well as to update the text to reflect changes in the most recent versions of SPSS/PC+. The description of the base system has been updated to deal with the new opening display and the various developments of REVIEW, and I have added information on two important new procedures, RANK and RELIABILITY. I have also extended, in the final chapter, the section which briefly describes procedures which are beyond the scope of this book. I have omitted the extensive coverage of REPORT, because I found that most students who attempted to use this procedure found it tedious, and because I now encourage students to take full advantage of listing files, and to produce reports by incorporating edited extracts of listing files in their word-processed report. Three new appendices have been added ('Notes for Trainers', 'Using Word-processor and Other Files with SPSS/PC+', and a 'Brief Guide to REVIEW Commands'), and the Glossary (*Appendix G*) has been extended.

Finally ...

In my early days of teaching SPSS/PC+, I tended to show off the exceptional power and flexibility of the SPSS/PC+ system by demonstrating some of its more remarkable capabilities and explaining how the system was working. The intention was not to mystify students, but to show that this was a system to be reckoned with, and a system worth learning. I remember that on one of these occasions a student remarked: 'I don't want to understand the system, I just want to analyze my results'. This is an instructive point. Despite initial apprehension and some antagonism, however, this student soon became delighted by SPSS/PC+, as do so many others. Not only are students 'grateful' to the system for providing them with the analyses they need, and for saving them a lot of time and effort, but they take pride in having mastered a system used by professional researchers. Familiarity with SPSS/PC+ also encourages students to explore new avenues of analysis and can thus broaden their statistical skills.

My own enthusiasm for teaching for SPSS/PC+ to students reflects the fact that a basic proficiency in using the system encourages students to explore their dataset fully and systematically. Once the data and data definition files have been written, analyses can be carried out rapidly. Compared with other ways of analyzing data, less time is needed to operate the system, and more time is available for planning analyses and for considering the meaning of statistical output. The listing file provides the means for producing well-presented reports without having to laboriously type or write a copy of the results of an analysis. Moreover, competence in using SPSS/PC+ is a 'portable skill'. Students who become familiar with the system at college may later employ their skills in advanced research and may include mention of their proficiency as a useful item in their *curriculum vitae*. I know of several cases in which graduates' familiarity with SPSS/PC+ gave them a significant advantage over other applicants in their bid for employment in the research field and in the commercial field.

SPSS/PC+ is used extensively in clinics, social work and probation offices, voluntary agencies and government departments, as well as in most academic institutions and in a host of commercial companies (including banks, building societies, motor manufacturers, breweries and large retail organizations). It is recognized by research funding bodies as offering the researcher excellent facilities. Originating in the United States, it is used worldwide.

As a trainer you can gain satisfaction from knowing that you are introducing your students to a system generally acknowledged to be the best of its kind, and one that is likely to be used by many of them for many years to come.

Appendix B: Using Word-processor and Other Files with SPSS/PC+

Some of the types of file that SPSS/PC+ uses are editable ASCII files, and others are not (see *Appendix H*). The SPSS/PC+ editor, REVIEW, can create and edit only ASCII files.

One way to find out whether a file is in ASCII form is to try to display it on the screen with the **DOS** type command, using the format:

```
C>type myfile
```

If all of the characters on the screen are alphanumeric, then the file is probably in ASCII form. If the screen displays characters which are not available on the computer keyboard, however, then the file is not in ASCII form.

The files normally produced by word-processors contain information other than that expected by SPSS/PC+ (escape sequences and control characters) which means that they are generally not acceptable to SPSS/PC+.

Similarly, files produced by spreadsheet and database programs contain information reflecting the way in which the particular program works as well as information about the dataset. This means that, like word-processed files, they cannot be entered directly into SPSS/PC+.

Fortunately there are several ways of overcoming these problems of compatibility, and this appendix will provide a general guide to the strategies by which 'foreign' files can be made acceptable to SPSS/PC+ and how files produced by SPSS/PC+ can be edited in a foreign environment (an editor program or a word-processor).

Creating files for use by SPSS/PC+

Files to be used by SPSS/PC+ can be created in several different foreign environments, by editor programs and word-processors, for example, and by spreadsheet and database management systems.

We will first consider the creation of editable text files. These may be data files (which contain only numbers), data definition files, or other command files. We will then consider the use of files created by spreadsheet and database programs.

Creating text files

There are many advantages of using the SPSS/PC+ editor, REVIEW, to create certain SPSS/PC+ files (the data definition file and other command files, or the data file, for example) during an SPSS/PC+ session. However, the process of writing long data files or data definition files can take a considerable time, and if the SPSS/PC+ facility is not within easy reach, or is not always available, it might be more convenient to write such files using another editor.

Many word-processing programs and editor programs are suitable. Files for use by SPSS/PC+ should be kept as 'plain' as possible, avoiding any special 'control' facilities which the package might offer. The files should consist of alphanumeric text set in lines of no more than 80 columns.

SPSS/PC+ recognizes the end of a line by 'carriage-return' information, but some word-processors with an automatic line-wrap facility do not automatically include the necessary end-of-line sequence. Thus in writing a file for use by SPSS/PC+ you may need to end each line with <RETURN>. A **<Ctrl-Z>** command should be used to end the file.

Many word-processors offer you the option of saving a file as an ASCII file, and several file-conversion programs are available which convert text files from the format of various word-processors to ASCII form (and vice-versa).

If you are familiar with a particular editor or word-processor you might like to try composing a simple data definition file to see whether the result proves acceptable to SPSS/PC+. If you encounter difficulties, try editing the file with REVIEW. Sometimes the removal of obvious 'blemishes' within the file will be sufficient to make the contents usable by the system.

You may be even able to write files on a computer which is not PC compatible. Several commercially available programs (*Media Master* is one example) are able to copy text files written in a foreign disk format to the standard PC format. Thus you could write the file using your preferred editor program or word-processor on a readily available computer which is not PC-compatible and then transfer the completed file to an PC format disk. This disk would then be inserted into either *Drive A* or *Drive B* of the PC and the file copied into the appropriate directory of *Drive C* (the hard disk). The SPSS/PC+ system would then access the file and process the information. Such files can also be re-edited with REVIEW.

It is not possible to include here details of the many editors which can be used to produce files for SPSS/PC+. However, we will briefly consider the use of *WordStar*, which differs in some ways from the majority of other popular word-processors.

Using WordStar
There are two basic rules for using *WordStar* to produce SPSS/PC+ files:

a) use the 'non-document' file mode
b) keep to text material (avoid 'controls')

a) When preparing a text file with *WordStar* the 'document' file mode is generally used. A new or existing file is edited by issuing a 'D' command. The 'document mode' sets a line length, provides an automatic word-wrap at the end of a line, and makes page breaks as appropriate. Such facilities, however, will produce errors when used to create a file for SPSS/PC+. Thus the alternative non-document mode is used. To create or edit a non-document file you enter the command 'N' (rather than 'D'). You may now type normally and use the normal editing controls, but there is no automatic return at the end of a line, and no automatic page break. Remember to keep all lines to a maximum of 80 columns, and issue <RETURN> at the end of each line.

b) When using *WordStar* to produce documents, several print commands may be used. Some are special instructions to the printer ('dot commands') and others are inserted within the text material to control such features as underlining and bold printing. Do *not* use commands of these types when preparing a file for use by SPSS/PC+.

The completed file – a matrix of numbers constituting a data file perhaps, or a set of statements constituting a data definition or other command file – can now be saved using the usual *WordStar* 'save' command.

If problems are encountered, the file can be re-edited using REVIEW. If, in error, you have used the 'document mode' to prepare a file for SPSS/PC+, it may be possible to make the file acceptable to SPSS/PC+ by editing out the 'page break' symbol and any other 'odd' symbols.

Creating other files

Files produced by spreadsheet and database management systems can also be transformed for use by SPSS/PC+, but the process is very different from that involved in producing suitable word-processed files. In using these systems, the 'foreign' file (which will include not only the data itself but also data definers and descriptors) is first produced in the usual way, using the normal spreadsheet or database commands, and this file is then translated by SPSS/PC+ to produce an active file. This file may then be saved as a system file. Unlike files produced by word-processors and editor programs, these files cannot be edited (for example, by REVIEW).

The following sequence describes one way of creating an SPSS/PC+ system file from a file produced by one of the popular spreadsheet or database programs (including *1–2–3*, *Symphony* and *dBase-IV*):

(i) copy the foreign file to your directory on the hard disk
(ii) use the appropriate SPSS/PC+ TRANSLATE command to translate the foreign file into the SPSS/PC+ active file
(iii) save the active file to disk (as a system file), using the SAVE command.

Two further points need to be added. The whole contents of the foreign file do not need to be included in the final system file. Variables can be excluded when TRANSLATE is used, or when SAVE is used, by including appropriate subcommands. Also, some translated files may be arranged such that the rows represent the variables and the columns the cases. This is the opposite of what SPSS/PC+ expects, and such a dataset needs to be 'turned through 90 degrees' using the FLIP command immediately after the TRANSLATE command.

Files from several of the more popular database and spreadsheet systems can be translated into SPSS/PC+ files using the TRANSLATE command, which forms part of the SPSS/PC+ base module. However, the optional *DBMS/COPY Plus* module translates between SPSS/PC+ system files and over 60 popular database, spreadsheet and statistical software packages. Files can be translated from the foreign package to SPSS/PC+ and vice versa, and selections of variables and cases can be made. Files translated with *DBMS/COPY Plus* retain their necessary dictionary information.

Editing files produced by SPSS/PC+

SPSS/PC+ automatically produces, for each session, a listing file which contains all of the output tables, graphs and statistics produced during the session, and a log file of all the commands issued to the system (together with warning messages and output page references). The listing file is named 'SPSS.LIS' and the log file is named 'SPSS.LOG' (unless either specification has been changed by the user). Both files are editable ASCII files.

The best way of obtaining a printed copy of the output is to work with an edited copy of the listing file (results can be printed *during* the session, but this tends to slow things up).

Both the log file and the listing files are 'inititialized' (deleted) at the beginning of the next SPSS/PC+ session to clear the files for the new log and listing information. The contents of files from successive sessions may be retained, however, if the names of the two files are changed immediately after an SPSS/PC+ session has ended. To do this, use the **DOS** rename command, for example:

```
C>ren spss.lis  lisoutl.doc
```

This command would rename the existing listing file (with the default name 'SPSS.LIS') as 'lisoutl.doc'.

When you want to edit such a file you could use REVIEW, but there are considerable advantages in using a word-processor. In particular, word-processors paginate the material in a form suitable for printing, they allow the insertion of special print commands (specifying that some of the text should be printed in bold or italics, for example), and they allow the user to specify the typeface and size of print to be used.

Reading the re-named listing file into your word-processor should present no problem. In some cases the program will only edit files named with a particular file extension (thus *Word* needs the extension .doc) and this should be taken into account when renaming files. The contents of the file should then be displayed on the screen.

Certain word-processors (including *WordStar* and *Word*) do not display exactly the same characters (for horizontal and vertical lines, for example) as the original ASCII file. To obtain a copy without the inappropriate characters you can edit the file manually, or use the word-processor facility for changing strings of characters throughout the text. With large files it may be worth using a file conversion program (*Star Exchange* is one example) to convert the file from ASCII to a form that is perfectly compatible with your word-processor.

SPSS/PC+ output is displayed across 79 columns, and this means that the output from some procedures (for example where large correlation matrices are displayed) may be too wide for you to print comfortably across a single page. You could remedy the problem by setting the 'pitch' on your printer to a higher number (this would compress the printing), or by editing the material (perhaps making used of the word-processor 'column move' facility) in order to arrange it on the page in a suitable form for printing.

You may then wish to develop your listing file into the results section of your project report, by annotating the output, deleting redundant material, labelling and numbering tables, adding commentary, and specifying which sections are to be printed using bold or underlined (or italic) text.

Appendix C: The Gender Attitudes Survey: Questionnaire

The questionnaire for the Gender Attitudes Survey was designed specifically to collect data that would allow many features of the SPSS/PC+ system to be illustrated. The 50 respondents were all psychology students from one British University. All students attending psychology classes on two particular days were asked to complete the questionnaire once. The 176 completed questionnaires were then numbered and a table of random numbers used to select 50 cases for inclusion in the sample. This limitation was made to reduce the dataset to a size which readers of this book could enter into a file without undue effort (it takes about three-quarters of an hour to create a file containing the complete set of data).

The questionnaire begins by asking for certain personal details (sex, age, faculty and year of study) and then asks the respondent to indicate the degree of agreement or disagreement with each of ten statements. These statements were derived from preliminary discussions about gender issues with a small group of students.

There are then ten 'first person' statements about emotions and interpersonal relationships, and the respondent is asked to judge whether more men or women would be likely to agree with each of these statements. Responses to these items may be interpreted as reflecting stereotypes about men and women. Later in the questionnaire (and over the page, to reduce the likelihood that respondents might refer back to their former responses) the same set of statements is presented. This time, however, the subject is asked whether the statement does or does not apply personally. By examining the proportions of males and females agreeing with each of these items, some indication is gained of whether any stereotypes which emerge actually reflect the personal responses of the young men and women who took part in the survey.

Following a question about marital status, unmarried people, only, are directed to a set of items designed to assess the importance attached to various characteristics for someone who would be a 'desirable date'. Again, these items are repeated over the page, and on the second occasion the respondent is asked to indicate the importance attached to the characteristics for someone who might be 'a desirable marriage partner'. A comparison between the two sets of responses might indicate that the importance of particular features differs somewhat for a desirable temporary and permanent partner.

Questions on weight, height and expected salary were included in order to provide data that (like age) could be regarded, non-controversially, as 'true numbers' rather than as codes, thus permitting analyses which demand that data be truly numerical.

The 12 questions from the short form of the Eysenck Personality Inventory (Eysenck, 1958) were included to enable two personality characteristics – extraversion and neuroticism – to be measured. These questions are not included in the questionnaire reproduced in this Appendix, but the relevant data are included in *Appendix E*.

G A SURVEY

PLEASE LEAVE BLANK

SEX (circle one) M / F AGE .... (YRS) | ID _ _ _ _

Indicate (by circling the appropriate number) your **Faculty** of study : | SEX _

AGE _ _

1 – ARTS 2 – SCIENCE 3 – SOCIAL STUDIES | FAC _

Year of study (circle one) First Second Third | YR _

Please circle the number beside each of the statements below to indicate how much you agree or disagree :

	Strongly Agree						Strongly Disagree	
1. Women are 'equal' to men in every way	1	2	3	4	5	6	7	VAR001 _
2. New legislation is needed to ensure equal rights	1	2	3	4	5	6	7	VAR002 _
3. Many women use 'sex discrimination' as an excuse for their own shortcomings	1	2	3	4	5	6	7	VAR003 _
4. There are many activities in which men are clearly superior to women	1	2	3	4	5	6	7	VAR004 _
5. The 'women's movement' has had little serious effect	1	2	3	4	5	6	7	VAR005 _
6. The legal system often treats men more harshly than women	1	2	3	4	5	6	7	VAR006 _
7. Women are emotionally more sensitive than men	1	2	3	4	5	6	7	VAR007 _
8. Generally speaking, men have a higher level of sexual desire than women	1	2	3	4	5	6	7	VAR008 _
9. Most men make a much less serious commitment to marriage than women	1	2	3	4	5	6	7	VAR009 _
10. Most men feel very threatened by highly intelligent women	1	2	3	4	5	6	7	VAR010 _

			PLEASE LEAVE BLANK
If a survey were conducted with a very large 'representative' sample of men and women (in equal numbers), do you think that more MEN or more WOMEN would AGREE with each of the following statements (circle MEN or WOMEN against each item):			
	(circle ONE)		
I often feel a little depressed	MEN	WOMEN	ST01 _
I seem to think about sex a lot of the time	MEN	WOMEN	ST02 _
I often feel a little anxious	MEN	WOMEN	ST03 _
I have to admit that I get angry quite often	MEN	WOMEN	ST04 _
I enjoy figuring out abstract problems	MEN	WOMEN	ST05 _
I enjoy taking risks	MEN	WOMEN	ST06 _
I frequently have trouble sleeping	MEN	WOMEN	ST07 _
I tend to try to dominate other people	MEN	WOMEN	ST08 _
I am often bored	MEN	WOMEN	ST09 _
I find it difficult to trust people	MEN	WOMEN	ST10 _

What is your current marital status? (tick one)

Single / Married / Divorced / Separated / Widowed — MARIT _

... ...

If you are **MARRIED** do NOT answer the following question but turn to the next page.

FOR UNMARRIED PEOPLE ONLY

What do you look for in a 'desirable date'? Using the list of attributes printed below, indicate the importance you would attach to each when thinking about dating a person.

V = Very important　　I = Important
N = Not important　　U = Undesirable

IT IS (V or I or N or U) that my DATE should have ....

Ambition	V	I	N	U	DAT01 _
Intelligence	V	I	N	U	DAT02 _
Physical attractiveness	V	I	N	U	DAT03 _
A sense of humour	V	I	N	U	DAT04 _
Kindness	V	I	N	U	DAT05 _
Sexual experience	V	I	N	U	DAT06 _
Good morals	V	I	N	U	DAT07 _
Dependability	V	I	N	U	DAT08 _

P.T.O.

			PLEASE LEAVE BLANK
Statements which you have considered previously are presented once more below. This time you should consider whether each of the statements applies to YOU yourself. Circle either TRUE or FALSE to indicate whether the statement is TRUE for you or FALSE for **YOU**.			
	(circle ONE)		
I often feel a little depressed	TRUE	FALSE	PRS01 _
I seem to think about sex a lot of the time	TRUE	FALSE	PRS02 _
I often feel a little anxious	TRUE	FALSE	PRS03 _
I have to admit that I get angry quite often	TRUE	FALSE	PRS04 _
I enjoy figuring out abstract problems	TRUE	FALSE	PRS05 _
I enjoy taking risks	TRUE	FALSE	PRS06 _
I frequently have trouble sleeping	TRUE	FALSE	PRS07 _
I tend to try to dominate other people	TRUE	FALSE	PRS08 _
I am often bored	TRUE	FALSE	PRS09 _
I find it difficult to trust people	TRUE	FALSE	PRS10 _

What would you look for in a 'desirable marriage partner'? Consider each of the attributes printed below, and indicate the importance you would attach to each when thinking about marrying someone.

V = Very important I = Important
N = Not important U = Undesirable

IT IS (V or I or N or U) that someone I would MARRY should have ....

Ambition	V	I	N	U	MAR01 _
Intelligence	V	I	N	U	MAR02 _
Physical attractiveness	V	I	N	U	MAR03 _
A sense of humour	V	I	N	U	MAR04 _
Kindness	V	I	N	U	MAR05 _
Sexual experience	V	I	N	U	MAR06 _
Good morals	V	I	N	U	MAR07 _
Dependability	V	I	N	U	MAR08 _

Now please answer the following questions :			HTF _ HTI _ _
What is your HEIGHT?	.... ft.	.... inches.	
			WST _ _
What is your WEIGHT?	.... st.	.... lbs.	WLB _ _
If you were to get a job immediately after completing your current degree, what salary would you **EXPECT** to get in your **first** year (i.e. over the first 12 months) – do NOT allow for inflation, but give the salary at present money values –			SALARY _ _ _ _ _ E1 _ E2 _ E3 _ E4 _
	£		E5 _ E6 _ E7 _ E8 _
NOW PLEASE COMPLETE THE LAST SET OF QUESTIONS, OVERLEAF			E9 _ E10_ E11_ E12_

Appendix D: The Gender Attitudes Survey: Codebook

The next pages contain a full codebook for the Gender Attitudes Survey, including the column number occupied by the variable (1–80), the 'type' of data (**N** = 'numerical' ; **C** = 'categorical'), the valid values and 'missing' value, and any value and variable labels.

Several additional variables (*neur*, *extra*, *sexism*, *sextract* and *problem*) were created from the original data set. *Appendix F* includes the **COMPUTE** commands used to create these variables. For some of the analyses included in this book, the values for the original data and for new variables were recoded using an appropriate **RECODE** command. See the text and *Appendix F* for further details.

Variable Name	**Col/s**	**'Type'**	**Values**	**Value Labels**	**Miss.**	**Var. label**
ID	1–4	N	–	–	9999	Subject No.
SEX	5	C	1 2	Male Female	9	Subject sex
AGE	6–7	N	–		99	Subject age
FAC	8	C	1 2 3	arts science soc. studies	9	Home faculty
YR	9	C	1 2 3	1st year 2nd year 3rd year	9	Year of study
VAR001	10	?	1 to 7	strongly agree / strongly disagree	9	Women equal to men
VAR002	11	?	1 to 7	strongly agree / strongly disagree	9	New legislation needed

Variable Name	Col/s	'Type'	Values	Value Labels	Miss.	Var. label
VAR003	12	?	1 to 7	strongly agree / strongly disagree	9	Women use sd as excuse
VAR004	13	?	1 to 7	strongly agree / strongly disagree		Men superior in many activities
VAR005	14	?	1 to 7	strongly agree / strongly disagree	9	Women's move. ineffective
VAR006	15	?	1 to 7	strongly agree / strongly disagree	9	Legal system anti-men
VAR007	16	?	1 to 7	strongly agree / strongly disagree	9	Women more sensitive
VAR008	17	?	1 to 7	strongly agree / strongly disagree	9	Men more sexy
VAR009	18	?	1 to 7	strongly agree / strongly disagree	9	Men less marriage commitment
VAR010	19	?	1 to 7	strongly agree / strongly disagree	9	Men threat. by intell. women
ST01	20	C	1 2	MEN WOMEN	9	Depressed often (M-W)
ST02	21	C	1 2	MEN WOMEN	9	Sex thoughts often (M-W)
ST03	22	C	1 2	MEN WOMEN	9	Anxious often (M-W)
ST04	23	C	1 2	MEN WOMEN	9	Angry often (M-W)

Variable Name	Col/s	'Type'	Values	Value Labels	Miss.	Var. label
ST05	24	C	1 2	MEN WOMEN	9	Enjoy abstract problems (M-W)
ST06	25	C	1 2	MEN WOMEN	9	Enjoy risks (M-W)
ST07	26	C	1 2	MEN WOMEN	9	Trouble sleeping (M-W)
ST08	27	C	1 2	MEN WOMEN	9	Domin. people (M-W)
ST09	28	C	1 2	MEN WOMEN	9	Bored often (M-W)
ST10	29	C	1 2	MEN WOMEN	9	Trust diffi-cult (M-W)
MARIT	30	C	1 2 3 4 5	Single Married Divorced Separated Widowed	9	Marital status
DAT01	31	C	1 2 3 4	Very Imp. Important Not imp. Undesirable	9	Ambition (date)
DAT02	32	C	As	DAT01	9	Intelligence (date)
DAT03	33	C	As	DAT01	9	Phys attract. (date)
DAT04	34	C	As	DAT01	9	Sense of hum. (date)
DAT05	35	C	As	DAT01	9	Kindness (date)
DAT06	36	C	As	DAT01	9	Sexual exper. (date)
DAT07	37	C	As	DAT01	9	Good morals (date)
DAT08	38	C	As	DAT01	9	Dependability (date)
PRS01	39	C	1 2	TRUE FALSE	9	Depressed often (SELF)
PRS02	40	C	1 2	TRUE FALSE	9	Sex thoughts often (SELF)

Variable Name	Col/s	'Type'	Values	Value Labels	Miss.	Var. label
PRS03	41	C	1 2	TRUE FALSE	9	Anxious often (SELF)
PRS04	42	C	1 2	TRUE FALSE	9	Angry often (SELF)
PRS05	43	C	1 2	TRUE FALSE	9	Enjoy abstract problems (SELF)
PRS06	44	C	1 2	TRUE FALSE	9	Enjoy risks (SELF)
PRS07	45	C	1 2	TRUE FALSE	9	Trouble sleeping (SELF)
PRS08	46	C	1 2	TRUE FALSE	9	Domin. people (SELF)
PRS09	47	C	1 2	TRUE FALSE	9	Bored often (SELF)
PRS10	48	C	1 2	TRUE FALSE	9	Trust diffi-cult (SELF)
MAR01	49	C	1 2 3 4	Very Imp. Important Not imp. Undesirable	9	Ambition (spouse)
MAR02	50	C	As	MAR01	9	Intelligence (spouse)
MAR03	51	C	As	MAR01	9	Phys attract. (spouse)
MAR04	52	C	As	MAR01	9	Sense of hum. (spouse)
MAR05	53	C	As	MAR01	9	Kindness (spouse)
MAR06	54	C	As	MAR01	9	Sexual exper. (spouse)
MAR07	55	C	As	MAR01	9	Good morals (spouse)
MAR08	56	C	As	MAR01	9	Dependability (spouse)
HTF	57	N		–	9	Height (feet)
HTI	58–59	N		–	99	Height (inches)
WST	60–61	N		–	99	Weight (stone)
WLB	62–63	N		–	99	Weight (pounds)

Variable Name	Col/s	'Type'	Values	Value Labels	Miss.	Var. label
SALARY	64–68	N		–	99999	Annual salary (£s sterling)
E1	69	C	1	YES	9	H/D no reason
			2	NOT SURE		
			3	NO		
E2	70	C		As E1	9	Prefer action
E3	71	C		As E1	9	Mood w/o cause
E4	72	C		As E1	9	Rapid action
E5	73	C		As E1	9	Moody
E6	74	C		As E1	9	Mind wander
E7	75	C		As E1	9	Make friends
E8	76	C		As E1	9	Quick-sure acts
E9	77	C		As E1	9	Lost in thought
E10	78	C		As E1	9	Lively
E11	79	C		As E1	9	Bubbly/sluggish
E12	80	C		As E1	9	Social contacts

Appendix E: The Gender Attitudes Survey: Data

The two pages of this Appendix contain *all* of the data from the Gender Attitudes Survey which has been subjected to analysis throughout this book. There are **50 cases**. The **69 variables** extend over 80 columns.

The 'guide-lines' printed after each segment of 10 columns are to help with data entry – they should not of course be entered with the data. Column numbers are also given above the data matrix, again to aid data entry (as you enter data, using **REVIEW**, a number at the bottom of the screen indicates the current cursor column).

(Gender Attitudes Survey Data – Cases 0001 to 0020)

```
         1 1111111112 2222222223 3333333334 4444444445 5555555556 6666666667 7777777778
1234567890 1234567890 1234567890 1234567890 1234567890 1234567890 1234567890 1234567890

0001120122|1636477732|1221122221|3121333322|2211121231|2123226011|1000600011|3313112133
0002220121|1347663622|1211121211|2221133122|2221222222|2113315091|0140520011|3231113211
0003120225|3435524542|1211121221|3211132311|1212121131|1113235111|0021700011|1111231213
0004219122|1476437312|2211111211|2222132111|1222222222|2213215050|8050750013|1231332111
0005226122|1446626312|1212121221|2221132211|1112212111|2112115051|1000650013|2311131121
0006220121|6755465542|9219921991|2221232112|1111121122|2113215020|8000800013|1111321121
0007120135|5635227762|1211122111|1111144121|1212222111|1113116001|1080700013|1113213323
0008120122|2235252652|1221121121|9999999921|1211222221|1113225111|0000650031|3133233313
0009219122|1524527452|1221111221|3231132211|1222221121|3112219990|8129999911|1312131113
0010224122|1612517652|2211111221|2211132111|1222221122|1113215091|1000900013|2131211111
0011227323|2727434432|1212112121|3222232322|2222222232|2223235020|8070700031|3131111121
0012122235|3336325332|1211111211|2211233322|2211112211|2223325101|4000750011|1111121111
0013219113|1476364642|1122212111|1121131112|1121211211|2113115000|8000400031|3131113111
0014245111|2456363562|1299999921|2121132222|1112221221|2113225061|1000500011|1323113113
0015122235|2436333322|1211121211|2111123121|2212212122|1112215101|3020900033|3312131131
0016231132|1345766631|9999999992|9999999922|2111222232|3223225021|0000650032|3131132313
0017220135|1666747762|9299199921|2121131212|1222222221|2223125040|9000650013|1231131111
0018222133|2536434319|9999999991|2111121112|1211222221|1112115010|9080970032|3131113121
0019221117|1655436532|1221121111|2221131112|1211222232|3113115080|8120750031|1331121331
0020124111|1773466772|1221122221|2121233211|1211121221|2123325101|1120700013|1111311313
```

(Gender Attitudes Survey Data – continued – Cases 0021 to 0050)

1	1111111112	2222222223	3333333334	4444444445	5555555556	6666666667	7777777778
1234567890	**1234567890**	**1234567890**	**1234567890**	**1234567890**	**1234567890**	**1234567890**	**1234567890**
0021221329	1666426412	2221111211	3221132211	2222221132	2113225091	2000600011	1331332331
0022219322	1777777732	1221121211	3231131111	1122112222	2113115081	1060750011	1111121211
0023229321	1437957752	1222212212	9999999922	1111212211	2113115020	8060700031	3131113231
0024118212	2342213622	1211111121	3222132122	1111291122	2113215091	1020700013	1113111123
0025221312	1445657422	1211121111	2211133122	1211222122	1113325060	9070700021	3131333111
0026148315	7636366432	1212122222	9999999911	1221121131	2224115081	1081000013	2111313333
0027121211	2563363732	1211121111	3111132121	2222222231	1113215111	2070600032	1331131112
0028231334	3336535532	1211129122	9999999912	1111221122	3222225050	8080750011	1111311211
0029120116	1323523342	1222222111	3321132221	2222222233	2223226001	2080400011	3331311111
0030223331	1273177312	1211121221	1311132212	1121111112	1112219990	9021000011	1111211111
0031119335	1412461412	1221111211	3222232211	1211221232	2223225101	1000550011	1111131211
0032221335	1775256632	1221122221	2121133122	1212221221	2112315090	9000700013	3131112111
0033233311	1366257422	1122221221	2222232212	1111122122	2223225060	9020700013	1111131123
0034120313	5222342242	1211121211	3211123221	2111212122	2112315101	1070750011	3111113111
0035121126	4126572412	1212121121	2221132211	1211121132	2113215111	2030850011	1311331113
0036221311	1775357732	1221121221	3222233211	1112221122	2223225040	9030700023	3111331311
0037222235	2325626622	1221112121	2231132112	1111111122	3113215050	8020750011	1111131112
0038132324	2549323432	1211121122	9999999922	1211222222	2213125061	0000800021	3133313131
0039237335	4245344652	2111221212	9999999922	2221221223	3113215041	0050600013	2131113113
0040127325	7331532312	1211121211	3222132112	2212222232	1213215101	0100800013	1111211131
0041221232	1645737632	1221121221	3211132211	1112212212	1113215040	9070720013	1311231311
0042121234	3335335552	1121121221	2222231211	1212222222	2223116011	4000700011	2111111231
0043221225	5436443642	1211111221	2131132222	2221222222	3123219990	9100800011	3131122113
0044120222	1725646662	1211111111	2221323221	2211222222	2123325071	0071200031	3121113131
0045119222	1532723442	1211121211	3222232212	2222122232	2223226001	3070700011	1323331111
0046221213	3545353642	1299921121	2221132211	1221222222	2113115040	8030750011	1311331332
0047122332	1655466442	1291121221	2221222321	2211211222	2122215070	9100700011	1131112111
0048220315	1623612542	1221121111	3111131111	1112122131	2113115061	1000500011	1111113211
0049120312	2655434632	9919991991	3211132111	1211221231	1113215111	2010700011	1131113111
0050126332	1552462562	1111121221	2232142212	1212222122	3223225091	2100800013	3231131213

Appendix F: The Gender Attitudes Survey: Data Definition File

"Gender.def"

This is a full copy of the data definition file used to specify to the system information concerning the variables in the "gender.dat" file (*Appendix G*). Despite its length the data definition file contains only four commands.

The DATA LIST command specifies the file where the data are to be found. It then names the variables in the order in which they appear within the dataset. Each variable name is followed by a number or number range indicating which columns the relevant data occupy. The TO convention is used a number of times to name a range of variables, and the format specification (column numbers) which follow each set will automatically be divided by the system into equal intervals, one interval for each variable.

The VARIABLE LABELS command provides labels for each of the variables. These labels will be included in the output of most procedures which make use of the variables.

The VALUE LABELS command provides labels for *some* of the values of *some* of the variables. Again this information will be included in the output of most procedures which make use of the variables.

The MISSING VALUES command specifies a value for each variable that is to be treated by the system as indicating that the relevant data are missing. These values will not be treated as numbers by the system (for example in calculating averages) but are taken by the system to indicate that no datum is available.

Only the DATA LIST command is essential. The labelling commands can be omitted. A MISSING VALUES command is needed whenever the dataset is incomplete (a common situation in social research).

```
DATA LIST FILE= "gender.dat" / id 1-4 sex 5 age 6-7 fac 8 yr 9
var001 to var010 10-19 st01 to st10 20-29 marit 30
dat01 to dat08 31-38 prs01 to prs10 39-48 mar01 to mar08 49-56
htf 57 hti 58-59 wst 60-61 wlb 62-63 salary 64-68
e1 to e12 69-80.
VARIABLE LABELS id "Subject No." / sex "Subject sex"
/ age "Subject age" / fac "Home faculty"
/ yr "Year of study" / var001 "Women equal to men"
/ var002 "New legislation needed"
/ var003 "Women use sexual dis. as excuse"
/ var004 "Men superior in many activities"
/ var005 "Women's movement ineffective"
/ var006 "Legal system anti-men"
/ var007 "Women more sensitive"
/ var008 "Men more sexy"
/ var009 "Men less marriage commitment"
/ var010 "Men threatened by intell. women"
/ st01   "Depressed often (M-W)"
/ st02   "Sex thoughts often (M-W)"
/ st03   "Anxious often (M-W)"
/ st04   "Angry often (M-W)"
/ st05   "Enjoy abstract problems (M-W)"
/ st06   "Enjoy risks (M-W)"
/ st07   "Trouble sleeping (M-W)"
/ st08   "Dominate people (M-W)"
/ st09   "Bored often (M-W)"
/ st10   "Trust difficult (M-W)"
/ marit  "Marital status"
/ dat01  "Ambition (date)" / dat02   "Intelligence (date)"
/ dat03  "Phys attract. (date)" / dat04 "Sense of hum. (date)"
/ dat05  "Kindness (date)" / dat06   "Sexual exper. (date)"
/ dat07  "Good morals (date)" / dat08   "Dependability (date)"
/ prs01  "Depressed often (SELF)"
/ prs02  "Sex thoughts often (SELF)"
/ prs03  "Anxious often (SELF)"
/ prs04  "Angry often (SELF)"
/ prs05  "Enjoy abstract problems (SELF)"
/ prs06  "Enjoy risks (SELF)"
/ prs07  "Trouble sleeping (SELF)"
/ prs08  "Dominate people (SELF)"
/ prs09  "Bored often (SELF)"
/ prs10  "Trust difficult (SELF)"
/ mar01  "Ambition (spouse)" / mar02  "Intelligence (spouse)"
```

```
/ mar03  "Phys attract. (spouse)"/ mar04 "Sense of hum.(spouse)"
/ mar05  "Kindness (spouse)" / mar06 "Sexual exper. (spouse)"
/ mar07  "Good morals (spouse)"/ mar08 "Dependability (spouse)"
/ htf    "Height (feet)"    / hti  "Height (inches)"
/ wst    "Weight (stone)"   / wlb  "Weight (pounds)"
/ salary  "Annual salary ## sterling"
/ E1      "H/D no reason"      / E2  "Prefer action"
/ E3      "Mood w/o cause"     / E4  "Rapid action"
/ E5      "Moody"              / E6  "Mind wander"
/ E7      "Make friends"       / E8  "Quick-sure acts"
/ E9      "Lost in thought"    / E10 "Lively"
/ E11     "Bubbly/sluggish"    / E12 "Social contacts".
VALUE LABELS sex 1 "Male" 2 "Female"
/ fac 1 "arts" 2 "science" 3 "soc. studies"
/ yr  1 "first year"  2 "second year" 3 "third year"
/ var001 to var010 1 "Strongly Agree" 7 "Strongly disagree"
/ st01 to st10 1 "MEN" 2 "WOMEN"
/ marit 1 "Single" 2 "Married" 3 "Divorced" 4 "Separated"
5 "Widowed"
/ dat01 to dat08 mar01 to mar08 1 "Very Imp." 2 "Important"
3 "Not Imp." 4 "Undesirable"
/ prs01 to prs10 1 "TRUE" 2 "FALSE"
/E1 to E12 1 "YES" 2 "NOT SURE" 3 "NO".
MISSING VALUES sex fac yr var001 to htf E1 to E12 (9)
/ age hti wst wlb (99) / id (9999) / salary (99999).
```

Transformed Variables

The following data transformation commands were used to **COMPUTE** new variables and to **RECODE** variables from the Gender Attitudes Survey. See Chapter 8 for general information on transformations. It will probably be most efficient to create a set of small command files, as below, and to use them when needed (e.g. for analyses in Chapters 11 and 12), rather than adding the commands into the data definition file. Some analyses use the original computed variables, and some use the variables after they have been recoded into categories. The relevant information is provided as examples are presented in the text.

EXTRANEU recodes personality items E1 to E12 and computes a neuroticism score (*neur*) and an extraversion score (*extra*).

```
recode e1 to e12 (2 = 0) (3 = -1).
compute neur = (e1 + e3 + e5 + e6 + e9 + e11).
compute extra = (e2 + e4 + e7 + e8 + e10 + e12).
```

RECEN recodes *neur* and *extra* scores into three categories (with approximately equal numbers in each).

```
recode extra neur (lo thru 0 = 1) (1 thru 3 = 2) (4 thru hi = 3).
```

SEXISM computes a *sexism* score from 7 of the 10 attitude items *var001* to *var010*.

```
compute sexism = (var001 + var002 + var009 + (8 - var003) +
    (8 - var004) + (8 - var005) + (8 - var008)).
```

RECSEX recodes *sexism* into three categories (with approximately equal numbers in each).

```
recode sexism (14 thru 20 = 1) (21 thru 26 = 2) (27 thru 38 = 3).
```

SEXTRACT – computes a *sextract* score from 6 of the 10 attitude items *var001* to *var010* (using the same formula as for *sexism* but omitting reference to *var003*).

```
compute sextract = (var001 + var002 + var009 + (8 - var004) +
    (8 - var005) + (8 - var008)).
```

RECSEXCT – recodes *sextract* into three categories (with approximately equal numbers in each).

```
recode sextract (10 thru 18 = 1)(19 thru 23 = 2)(24 thru 35 = 3).
```

SEXISM2 – computes a *sexism2* score from 7 of the 10 attitude items *var001* to *var010*, with the scoring for five of the variables reversed (*see* Chapter 14).

```
compute var003r = (8 - var003).
compute var004r = (8 - var004).
compute var005r = (8 - var005).
compute var008r = (8 - var008).
compute var009r = (8 - var009).
compute sexism2 = var001 + var002 + var003r
+ var004r + var005r + var008r + var009r.
```

PROBLEM – computes a *problem* score from 7 of the 10 personal problem items *prs01* to *prs10*.

```
compute problem = (2 - prs01) + (2 - prs03) + (2 - prs04) +
  (2 - prs07) + (2 - prs08) + (2 - prs09) + (2 - prs10).
```

Appendix G: Glossary of Statistical, Computing and SPSS/PC+ Terms

SPSS/PC+ keywords and commands (including subcommands) are printed in bold type.

ABSOLUTE VALUE: the value of a number or expression which ignores whether the original value is positive or negative. Thus the absolute value of -34 (and also of +34) is 34.

ACTIVE FILE: an SPSS/PC+ file, with data and dictionary information, which is currently in use. Certain changes (transformations, selections, etc.) may have been made since the file was originally made active and these changes may remain in effect until the end of the session. Thus the term active file refers to the file in use in its current state. The active file is a file in the computer memory, rather than a file stored on a disk, but the contents of the active file may be written to the disk as a system file by using a SAVE command.

ADVANCED STATISTICS: the name of an optional add-on enhancement for SPSS/PC+. It provides procedures for a number of advanced methods of analysis, including (in version 4.0) discriminant analysis, logistic regression, non-linear regression, log-linear modelling, and multivariate analysis of variance. The use of this package and these techniques are not described in this book. However, the methods of data entry, file creation, transformation, etc. applicable to the *SPSS/PC+ Advanced Statistics* module are the same as those used for the SPSS/PC+ base system.

ALL: an SPSS/PC+ keyword which can sometimes be used, instead of a list, to request that all variables be analyzed, or all statistics performed.

ALPHA COEFFICIENT: a measure of the reliability of a scale, based on the relationship between items within the scale. Alpha is mathematically equivalent to the average of the reliability coefficients that would be obtained from all possible split-halves. It is an estimate of the correlation between the scores actually obtained on the scale and a hypothetical 'true score'. The alpha coefficient can be requested within a RELIABILITY command. *See* Chapter 14.

ALPHANUMERIC VARIABLE: (also referred to as a string variable) a variable for which the values are stored as characters (which may include letters, symbols and numerals). Because the values are in character (rather than number) form, only a limited number of operations (including sorting) can be performed with alphanumeric variables. They cannot be subjected to arithmetic operations. *See also* long string; short string.

ALTERNATE FORMS RELIABILITY: reliability assessed by creating two parallel forms of a test, giving subjects both tests, and then correlating the scores obtained from both forms. *See* Chapter 14.

ANALYSIS OF VARIANCE: an inferential statistical procedure which allows comparison of the means (for a dependent variable) of groups defined by the values of one or more independent variables. If a single independent variable is involved then the SPSS/PC+ procedure ONEWAY may be used. If more than one independent variable is involved then ANOVA is the appropriate procedure. *See* Chapter 12.

AND: a logical operator used to join two or more relations, for example in a conditional transformation. Thus: IF (VAR1 EQ 2 AND VAR2 EQ 3) ....

ANOVA: an SPSS/PC+ command which performs analysis of variance.

AREA SAMPLING: a method of sampling in which an area is taken and divided into a number of smaller areas, or blocks. A sample of these blocks is then randomly selected, and samples are taken only from the selected blocks.

ARITHMETIC OPERATORS: the signs which give instructions for arithmetic operations to be carried out. The four basic arithmetic operations are addition (+), subtraction (-), multiplication (*) and division (/). SPSS/PC+ also allows exponentiation to be directly specified with '**'. Thus '2**3' would be read as: '2 to the power of 3', or '2 cubed' (i.e. 8). With these basic operations (and with brackets) complex transformations may be specified (for example, in a COMPUTE command). *See also* functions.

ASCENDING RANKING: assigning the highest rank, rank 1, to the case with the lowest value for the variable on which the rank-ordering is based, rank 2 to the case with the next lowest value, and so on. Thus rank-ordering people by their weight, in an ascending ranking, rank 1 would be assigned to the person with the lowest weight and the last rank position would be assigned to the heaviest person.

ASCENDING SORT: a sorting of cases within the current active file such that the case with the lowest value for a specified variable (or for specified variables) is placed at the beginning of the file, and the case with the highest value is placed at the end of the file. Thus in an ascending sort by age the youngest cases would appear at the beginning of the file and the oldest at the end. *See also* descending sort.

ASCII: The 'American Standard Code for Information Interchange' – a standard code which converts alphabetic and numeric information, and certain symbols and control characters, into binary code. Information coded according to this standard, and in which lines are terminated by a 'return' and the file is terminated with a 'Control Z', are 'text files' and can be written and read by a number of editors and by SPSS/PC+.

AUTOEXEC.BAT: a DOS batch file which is *auto*matically *exec*uted when DOS is loaded (or 'booted').

BACKUP: A copy of a file, a whole directory of files, or a whole disk made for safe keeping. It is always advisable to keep a spare diskette copy of files in case the hard disk copies are accidentally destroyed.

BARCHART: an SPSS/PC+ subcommand (of the FREQUENCIES command) which requests that the frequencies of data be plotted as a bar chart.

BAR CHART: a diagrammatic representation of data in which frequencies for a variable with relatively few distinct values are plotted graphically. *See also* histogram.

BATCH FILE: *See* batch mode.

BATCH MODE: a system whereby data or instructions are first prepared and then processed as a 'batch' by the system. SPSS/PC+ command files which have been prepared using an editor are batch files. Information given to the system in response to the SPSS/PC: prompt, however, is said to be entered in 'interactive mode'. In DOS a batch file is a file which contains one or more DOS commands. It is commonly given the file extension .BAT. Sometimes a batch file is created which will be *auto*matically *exec*uted when DOS is loaded (or 'booted'). Such a batch file is named 'autoexec.bat'.

BEEP: the sound issued by the system to signal an error (low-pitched sound) or the availability of the next screen of output (high-pitched sound). The beep can be switched off by the SET command: 'set beep off.'

BINARY CODE: a system for coding information which uses only the digits 0 and 1.

BINARY SITUATION: a situation (e.g. experiment, test, or question) in which there are only two possible outcomes (e.g. the toss of a coin, or a question to which the only permitted answers are 'Yes' or 'No').

BINOMIAL: an SPSS/PC+ subcommand (of the NPAR command) which requests a 'binomial test.' *See* Chapter 13.

BIT: A binary digit. The smallest unit used by a computer. Any decimal number between 0 and 255 can be represented by an 8-bit binary number (a 'byte').

BOXPLOT: a graphical plot of summary statistics for a frequency distribution. It plots each of the three quartiles (the 25th percentile and the 75th percentile provide the upper and lower boundaries of a box drawn in the display) and gives information about extreme values. A boxplot display can be requested within an EXAMINE command.

BUFFER: a small part of the computer's memory which acts as a temporary information storage space. For SPSS/PC+ to run efficiently the number of available buffers should be set to at least 8. If a suitable CONFIG.SYS file has been written, this will be done automatically when the machine is switched on.

BY: an SPSS/PC+ keyword. Used, for example, in a CROSSTABS command, e.g:'crosstabs sex BY soclass.'

BYTE: A group of eight bits, which together make a unit used to represent a character or number. The size of files, disks, diskettes and computer drives may be represented in terms of kilobytes (1 k = 1024 bytes) or megabytes (1 Mb = 1024k, just over one million bytes).

CASE: in a survey, a 'case' would usually represent one respondent. In a psychological experiment a case might represent one experimental subject (i.e. one person). A data set generally consists of the values relating to a standard set of variables for a number of cases. A case need not represent an individual person – it might represent a country, corporation, product, etc.

$CASENUM: a system variable which is automatically included for each case in the active file. The $ symbol at the beginning of the name indicates that the variable is a system variable rather than a user-declared variable. The value of *$CASENUM* is the position of that case in the dataset as the file is made active. If the order of cases is later changed by a SORT command, or if the size of the active file is reduced, the value of *$CASENUM* remains unchanged. Thus the original order of the initial active file (or what remains of it) can be re-established with the command: 'SORT by $CASENUM.' If the active file is saved as a system file, the *$CASENUM* assigned when the file was made active is included within the system file. *$CASENUM* is not, however, included in a portable file created with EXPORT.

CATEGORICAL VARIABLE: a variable for which all values conform to particular categories. Only a limited range of arithmetic procedures are permitted with such variables. *See also* code.

CELL: one intersection in a crosstabulation. *See also* chi-square test.

CENTRAL TENDENCY MEASURES: various statistics which identify different kinds of mid-point in a frequency distribution. Commonly, the mean, median, and mode.

CHARACTER: a letter, digit from 0 to 9, or any other recognized symbol or punctuation mark. The standard character set used in most computing is the ASCII set.

CHISQUARE: an SPSS/PC+ subcommand (of the NPAR command) which requests a 'One sample chi-square' test. *See* Chapter 13. Note also that the chi-square test can be requested in a Statistics subcommand within a CROSSTABS command.

CHI-SQUARE TEST: a statistical test which determines whether an observed pattern of results differs significantly from that which would be expected to occur merely by chance. Thus if it were assumed that by chance an equal number of boys and of girls would have blue eyes (rather than brown eyes) we could count the number of boys and girls with the two types of eye colour and subject the results to a chi-square test. A *significant* result would indicate that for our sample sex and eye colour were related – that the distribution of eye colour between the sexes was *not* random (or the result of mere chance). This would be an example of a 'two by two' chi-square (here, two sexes and two eye colours). The frequencies for each sex and each eye colour could be displayed in a crosstabulation with four cells.

CLOSED-ENDED QUESTION: a question for which a number of fixed alternative answers are provided. Examples include 7-point rating scales, Likert scales and other forms of multiple choice. By limiting valid responses to a fixed range of answers the data obtained are easily codable and readily amenable to SPSS/PC+ analysis.

CLUSTER ANALYSIS: A type of multivariate analysis which identifies groups (clusters) of similar cases. Similarity or closeness is judged on the basis of the values of cases for a set of variables. Two SPSS/PC+ procedures from the *SPSS/PC+ Statistics 4.0* module, CLUSTER and QUICK CLUSTER, perform cluster analyses.

COCHRAN: an SPSS/PC+ subcommand (of the NPAR command) which requests a 'Cochran test.' *See* Chapter 13.

CODE: a category label. To 'code data' is to categorise it. Some category systems reflect an underlying dimension – 'tall', 'average height', 'short' – and some do not – 'blue-eyed' 'brown-eyed'. The labels may be numbers (1 for 'blue-eyed', 2 for 'brown-eyed) but this should not be taken to imply that the codes express ranks or quantities. The term 'coding data' is often used to describe the process of creating the complete data matrix from a study, a process that might involve codes and real numbers.

CODEBOOK: a handy list, written by the user in the early stages of preparing a project, which contains the names of the variables to be included in a file, with information about the location, values, missing value, coding system, and labels for each variable. The codebook is not a computer file used by SPSS/PC+ but is written on paper for the user to consult. *See* Chapter 2. *See also Appendix D* for a full example of a codebook.

CODING: the process of translating raw data into categorised data.

CODING SYSTEM: any system by which raw data are categorized for analysis. Many coding systems are simple (for example, we could code people's ages into three categories – under 20; 20 to 40; over 40). Some are more complex – marital status, for example, may be coded using categories which take into account the individual's marital history ('twice widowed', 'twice married') as well as their present marital status. Very elaborate coding systems may need to be devised (usually after the data has been collected) if open-ended questions have been used.

COEFFICIENT: A statistic which provides a summary of a relationship. Thus a correlation coefficient provides a value for the degree of association between (usually two) variables.

COVARIANCE: a measure of the degree to which two variables covary. For two variables, X and Y, the covariance is measured by finding the average for X and Y, and then calculating each score as a deviation from the corresponding mean. Pairs of X and Y deviation scores are multiplied together. The sum of these products, divided by the number of cases, is the covariance.

COMMAND: an instruction issued to the SPSS/PC+ system either in batch mode or interactively. Operation commands display or change the current operating characteristics of the system. Data definition and transformation commands instruct the system on where and how data should be read, how data should be transformed and selected, and how output should be labelled. Procedure commands cause the data values to be read and, usually, instruct the system to perform a statistical analysis.

Procedure commands may include one or more sub-commands. Commands begin with a command word (or a three character truncation of the command word) and must end with a period (.).

COMMAND FILE: an SPSS/PC+ file containing a number of commands regarding procedures, transformations, selections, etc. This file is submitted to the SPSS/PC+ system as a batch file (using the INCLUDE command). A data definition file is one example of a command file, and the names of such files commonly include the extension .DEF. The names of other command files often include the extension .SPS.

COMPUTE: an SPSS/PC+ command which creates a new variable from existing variables (or modifies the values of an existing variable). It takes the form: 'compute C = A + B.' Complex compute commands are possible and may include several arithmetic operators, functions, relational operators and logical operators. The creation (or modification) of each variable needs a separate compute command.

CONDITIONAL TRANSFORMATION: an SPSS/PC+ data transformation which depends on one or more logical conditions being fulfilled. *See* logical expression; *see also* Chapter 8.

CONFIG.SYS: a DOS file which sets certain parameters to 'configure' or 'customize' the system for a particular type of use. Instructions contained within a CONFIG.SYS file are automatically executed when DOS is loaded (or 'booted').

CONTINGENCY TABLE: a tabulation of the joint frequencies for two or more variables. *See* crosstabulation.

CONTINUOUS VARIABLE: a numerical variable for which the values need not be integers. Thus height can be represented with decimal point data (e.g. 163.54) whereas an integer variable like 'number of children' must be represented by a whole number (1, 2, 3, etc.). Many of the data to be analyzed by SPSS/PC+ will be coded in number form but will be categorical rather than numerical. Because they are not numerical, coded variables should not be treated as either continuous variables or as true integer variables.

CORRELATION: an SPSS/PC+ command which causes a product-moment correlation to be performed.

CORRELATION: the correspondence between two sets of paired measurements. The numerical index used to indicate the degree of this correspondence is the correlation coefficient. The most common measure is the Product Moment Correlation (symbolized by r) based on a formula devised by Karl Pearson. The value of a correlation coefficient ranges from -1 (indicating total negative correlation) through 0 (indicating no association between the two sets of data) to +1 (indicating complete correspondence). A high positive correlation does not imply that the mean (i.e. average) values for each of the two variables are equal or similar but implies that high values of one variable are associated with high values of the other. A multiple correlation is a measure of the degree of relationship between a single dependent variable and a number of independent variables. *See also* correlation matrix, partial correlation.

CORRELATION MATRIX: a table of results of correlational analyses in which the correlation coefficients between each possible pair of a number of sets of data are represented.

CROSSTABS: an SPSS/PC+ procedure which produces a crosstabulation. The command takes the form: 'crosstabs sex by soclass.'; values of the first named variable determine the rows of the table, and values of the second named variable determine the columns. Various OPTIONS affect the display and the treatment of missing data, and various STATISTICS are available, including chi-square.

CROSSTABULATION: (or contingency table) a table showing the joint distribution of frequencies of cases on two or more discrete variables. Thus the frequency distribution for two sexes and three hair colours will be represented in a crosstabulation containing six cells. Crosstabulation is requested by using the SPSS/PC+ CROSSTABS command.

CURSOR: a lighted square or line on the computer screen which marks the current position at which information can be written.

CURVILINEAR RELATIONSHIP: a relationship between two variables which is shown graphically not as a straight line but as a curve.

DATA: numbers, measurements or codings. Raw data are data as they are collected, not having been processed or coded. The singular of 'data' is 'datum'.

DATA CASE: all of the data values relating to a single case.

DATA CLEANING: identifying and correcting errors in coding and data entry.

DATA RECORD: a collection of data that is only one line long.

DATABASE: an organized collection of data. *See also* database management system.

DATABASE MANAGEMENT SYSTEM: a program which organizes and manipulates data to produce summaries, comparisons, lists, etc.

DATA DEFINITION FILE: a command file which includes the data dictionary, i.e. information about the variables, their format, variable and value labels, missing data information, etc. The only essential command within a data definition file is the DATA LIST command which declares variables to the system. Data definition files include a reference to where the data may be found (e.g. in a data file or matrix file). It is possible to include the data within the data definition file itself, but this is not recommended.

DATA ENTRY II: an optional module of the SPSS/PC+ system which provides a way of entering, cleaning and editing data. Effort is needed to set up the initial commands which inform the system of certain parameters of the data, and the data file created cannot be edited as readily as data files created with REVIEW. DATA ENTRY II is particularly useful when a large and complex dataset is being entered. The use of this optional module is not covered in this book.

DATA FILE: a text file which contains only data.

DATA LIST: an SPSS/PC+ data definition command which specifies the names of variables and their position in the dataset. The data list specifies the name of a data file from which the data are to be read with a FILE subcommand that takes the form: DATA LIST FILE = "gender.dat" (followed by the list of variables and their location). By default, fixed format is assumed, but free format can be specified by: DATA LIST FREE (or, where a data file is to be read: DATA LIST FILE = "gender.dat" FREE, followed by the list of variables). A DATA LIST command may also refer to data written in a matrix file or other results file.

DATASET: a set of data. Usually comprises (in the case of SPSS/PC+ datasets) data for a number of variables from each of a number of cases.

DATUM: the singular of 'data'; one item within the dataset.

DECIMAL: the numbering system that uses the numerals 0 to 9.

DEFAULT: a value, option, name or instruction which is assumed by the system when no specification has been provided by the user. A passive default comes into operation without any action on the part of the user (and most defaults are of this type). An active default, however, requires the user to request that a default should be brought into operation. Thus if a STATISTICS subcommand within a FREQUENCIES procedure specifies 'DEFAULT', four default statistics are displayed. With no STATISTICS subcommand, however, no statistics are displayed.

DEGREES OF FREEDOM: a statistical term which refers to the number of values within a data set which are free to vary. For example, if we know that the sum of the ages of 6 children is 40 then knowing the ages of any 5 of the children will enable us to calculate the age of the sixth child. Thus in this case there would be five degrees of freedom. With a single restriction, as in this case, the value for the degrees of freedom (D.F) is 'N − 1' (in this case 6–1 = 5). D.F. is included as an element in many statistical formulae and is also used in determining the probability associated with the value of the statistic obtained. SPSS/PC+ assesses D.F. automatically, and uses it both in calculating the relevant statistic and the associated probability. The D.F. value is also displayed in the output from many procedures.

DEMOGRAPHIC: demography is the study of (usually human) populations. Demographic data include information on such variables as age, sex, income, social class and marital status. Such information is often collected in surveys on attitudes, etc. so that analyses relevant to these major population characteristics can be performed. This is the case even when it is not the major purpose of the survey to focus on such aspects.

DEPENDENT VARIABLE: a variable which is measured in two or more conditions. In an experiment, an independent variable (for example, level of noise) is manipulated and a dependent variable (for example, accuracy of word recognition) is measured. The term is also applied in other types of study. Thus in an attitude survey the sex of the respondent may be treated as an independent variable and a particular attitude as a dependent variable.

DESCENDING RANKING: assigning the highest rank, rank 1, to the case with the highest value for the variable on which the rank-ordering is based. Rank 2 is assigned to the case with the next highest value, and so on. Thus rank-ordering people by their weight, in a descending ranking, rank 1 would be assigned to the heaviest person and the last rank position would be assigned to the person who weighed least.

DESCENDING SORT: a sorting of cases within the current active file such that the case with the highest value for a specified variable (or for specified variables) is placed at the beginning of the file, and the case with the lowest value is placed at the end of the file. Thus in an descending sort by age the oldest cases would appear at the beginning of the file and the youngest at the end. *See also* ascending sort.

DESCRIPTIVE STATISTICS: an area of statistical analysis which describes data (their frequencies, averages, 'spread', correlations, etc.). This contrasts with inferential statistics. *See also* inferential statistics.

DICTIONARY: information, within a SPSS/PC+ file, about the variables referred to in that file. The dictionary can contain, besides the names of the variables, indications about their format and location within the data set, missing values, and both variable labels and value labels.

DIRECTORY: a logical (rather than physical) part of a hard disk (and, sometimes, a diskette). Because a hard disk may store many hundreds of files, some logical structure is necessary to allow efficient access by the user. Large capacity hard disks have several drives; smaller capacity hard disks have just one drive, Drive C. Each drive contains a root directory and is likely to have many sub-directories (often one for each of the major users). There may also be sub-sub-directories, so that the overall directory structure is tree-like. From within any one directory, access to other directories may be gained by issuing a 'change directory' command. Often path commands are included in the autoexec.bat command which is activated whenever the computer is switched on. Path commands allow information within one directory to be automatically accessed from another directory. Consult the user manual for the DOS system for further information.

DISCRETE VARIABLE: a variable which can take one of only a restricted set of values from within the overall range. The most common form of discrete variable is the integer variable in which only whole numbers can be appropriately used. Thus any answer to the question 'How many children are there in the family?' must be a whole number (0, 1, 2, 3, etc.). This variable is thus a discrete (or integer) variable.

DISCRIMINANT ANALYSIS: a statistical technique designed to estimate the relationship between predictor variables and one or more categorical response variables. This form of analysis is requested by the DSCRIMINANT procedure which forms part of the *SPSS/PC+ Advanced Statistics 4.0* module.

DISK DRIVE: a part of the computer hardware. Its function is to read information from and write information to a diskette or hard disk. *See* drive; Drive A.

DISKETTE: a flexible disk (usually 5.25 or 3.5 inches in diameter) which can be used in Drive A or Drive B of a microcomputer. Typically they are able to store 360k, 720k,

1.2Mb or 1.44Mb of information (k = 1 kilobyte; Mb = 1 megabyte). A diskette is also sometimes referred to as a 'floppy disk.' Diskettes should not be inserted into the appropriate drive until the computer has been switched on, has finished its initial routines and displays the opening prompt.

DISPLAY: an SPSS/PC+ operation command which lists the variables currently defined to the system. The command DISPLAY, without further specification, lists the names of the variables and any labels which have been provided for them. If the command specifies one or more variables by name, e.g.: DISPLAY sex age var001, then additional information (format, missing value and value labels) is displayed for each variable specified.

DOS: abbreviation for Disk Operating System. The set of programs which manages file manipulation, directory organization and the interaction between the computer processor and the monitor, keyboard, printer, etc. SPSS/PC+ uses MS-DOS.

DRIVE: a computer disk drive; a part of the computer hardware which can read information from and write information to a disk. Drive A and Drive B are drives which deal with removable diskettes. The first drive on the hard disk is referred to as Drive C, and large capacity hard disks may also have Drive D, Drive E, etc.

DRIVE A: a computer disk drive; a part of the computer hardware which can read information from and write information to a diskette. Drive A (and Drive B, if present) are accessed by slots mounted on the front of the main computer console. Drive A (and Drive B) may accommodate floppy disks of 5.25 or 3.5 inches in diameter, and are capable of writing information to a 360k, 720k, 1.2Mb or 1.44Mb format, depending on the Drive specification. Diskettes should not be inserted into either Drive A or Drive B until the computer has been switched on, has finished its initial routines and displays the opening prompt. *See also* disk drive; diskette.

DRIVE B: a computer disk drive; a second drive capable of reading information from and writing information to a diskette. Not all microcomputers have twin drives. Some only have one (Drive A). *See* Drive A.

DRIVE C: The first (or only) drive on the hard disk. Typically has 32Mb capacity. There may be other drives on the hard disk (Drive D, Drive E, etc.).

DROP: an SPSS/PC+ subcommand that can be used with the commands IMPORT and EXPORT, SAVE and GET, and JOIN. It is followed by a list of variables and instructs the system to exclude the named variables when executing the particular operation. Thus with an active file of 100 variables, if the SAVE command includes a DROP subcommand listing 10 of these variables, the saved system file will contain only 90 variables. When DROP is used, the variables in the new file remain in the same order as they were in the original file. *See also* keep.

ECHO: normally, commands issued to the system by the user are shown on the screen and included within the listing file. The SET command: SET echo off., however, will cause only the results of statistical and reporting procedures to be sent to listing and printer files.

EDITOR: a program which allows text to be written and edited. SPSS/PC+ has its own editor, REVIEW, but many other editors, including word-processing programs, database management systems, etc. can be used to create files for use by SPSS/PC+ and to edit the listing file produced by the system during each session.

ELSE : a keyword which can be used within a RECODE command, e.g.: RECODE VAR1 (0,1,2 = 1) (3,4,5 = 2) (ELSE = 3). ELSE recodes *all* other values including user-missing and system-missing values.

EQ: a relational operator meaning 'equal to'. Thus IF (VAR1 EQ 3) means 'if the value of VAR1 is equal to 3'. The equals symbol (=) can be used to replace EQ.

EQUAL LENGTH SPEARMAN-BROWN COEFFICIENT: an estimate of the reliability of a whole test, based on a split-half reliability when the two halves are of equal size. The formula used to derive this estimate assumes that the two halves of the test are equally reliable and have similar variance. The coefficient is included in the output from the RELIABILITY procedure when the split-half model is specified. *See also* split-half reliability; unequal length Spearman-Brown coefficient; Guttman split-half coefficient. *See* Chapter 14.

ERROR: (a) a problem diagnosed by the system. When the system diagnoses a mistake or inconsistency it will display an error message and will not execute a procedure. Amongst the most common errors are the incorrect spelling of a command word, blank lines inserted within a multi-line command, and a terminator period omitted from the end of a command. Mistakes in the declaration of variable and value labels, however, and most mistakes in data entry, will not be identified by the system.

(b) unexplained variability in measurement. Thus the score obtained on a test partly reflects the 'true score' and partly reflects error. Reliable scores are those for which the error component is low and the 'true score' component is relatively high.

EXPERIMENTAL RESEARCH: research aimed at testing causal explanations. Generally, one or more independent conditions are manipulated to observe the effect on one or more dependent variables.

EXPORT: an SPSS/PC+ command which produces a portable ASCII file from the current active file. Portable files include both data and data dictionary information as well as certain system information. They can be transported to other computers using KERMIT and are 'imported' into the SPSS system of the other machine using the IMPORT command.

F RATIO: an inferential statistical test calculated, for example, in an analysis of variance. In that case it is a ratio of the variability in the data produced by one factor (the numerator) to the overall variability produced by error (the denominator). More generally, a ratio of two variance measures obtained by dividing the larger of the values by the smaller.

FACTOR ANALYSIS: a complex statistical technique which computes, from a correlation matrix, the minimum number of 'factors' which will account for the overall pattern of inter-correlations. It provides a synopsis of the information contained in such a pattern. Factor analysis is used, for example, to identify the basic factors being measured by a large number of items within a personality inventory and may also be used to derive a suitable scoring method for such an instrument. Factor analysis is performed by using the FACTOR procedure included within the *SPSS/PC+ Statistics 4.0* module. The use of FACTOR is not considered in this book.

FACTORIAL DESIGN: a design which includes more than one independent variable.

FILE: a collection of information. Within the SPSS/PC+ system there are many types of file. They vary in terms of their content, their purpose and the form in which information is stored. Thus, in terms of content, some SPSS/PC+ files contain only data, others include data definition information, and some contain SPSS/PC+ commands. Others contain results of analyses or a record of commands entered during a session. For some types of file, information is stored in ASCII form. System files, however, are stored in binary form. The texts of some files are written by the user (using an editor.) Other files are written by the system at the request of the user, and some are written automatically by the system. *Appendix H* provides a table of the characteristics of the different types of file. *See also* active file; command file; data file; data definition file; listing file; log file; portable file; results file; system file.

FILE NAME: the name of a file, consisting of up to 8 characters and an extension of three letters following a period. Most file names are specified by the user, but for some files the system provides default names (e.g. 'SPSS.LIS' for the listing file and 'SPSS.LOG' for the log file). Although most file names do not *need* to include an extension, it is often useful to provide this. For example, .DEF may be used as the extension for a data definition file, .DAT for a data file, .SPS for a command file, and .MAT for a matrix results file.

FINISH: an SPSS/PC+ command which terminates the current SPSS/PC+ session and returns the user to DOS.

FIXED FORMAT: data are commonly organized within the dataset by placing data for the same variable for each case within precisely the same column(s), thus forming a perfectly rectangular data matrix. The system can then be informed of the position occupied by each variable. Fixed format is the default format used by SPSS/PC+ and it contrasts with free format in which successive variables are separated by one or more blank spaces.

FLOPPY DISK: *See* diskette.

FORCED CHOICE FORMAT: a response format in which mutually exclusive and discrete options are presented for the respondent to choose between.

FORMAT :

(a) the form, and position within the database, of data for each of the variables. A format specification for the data relating to a particular variable in an SPSS/PC+ file may include information about the location of the variable (i.e. columns occupied), the position of an implied decimal point (*See* Chapter 16) and the numeric or alphanumeric nature of the variable. *See also* fixed format; free format.

(b) to format a disk is to prepare a new disk so that it will be compatible with the computer system being used. Thus new diskettes to be used in Drive A will have to be formatted appropriately before files can be written to them.

(c) generally, the term 'format' describes the position and layout of a display or item. SPSS/PC+ allows the user a high level of control over format, often by means of a FORMAT subcommand.

FORMAT: a subcommand used within a number of procedure commands to specify how lists, plots, and output displays should be printed or displayed or how a results file should be written.

FORMATS: an SPSS/PC+ command which allows the user to change the way in which numeric variables are displayed and printed. The formats specified on such a command remain in effect for the rest of the session unless a new FORMATS specification is provided.

FREE FORMAT: a way of writing a data line, in which successive variables are separated by one or more blank spaces (and/or commas). This contrasts with the more usual fixed format method (the SPSS/PC+ default). When free format is used there is no need to add leading zeros to even up the space occupied by variables (as when 003 is entered for 3) and thus the data for different cases may be of different lengths. There is no insistence on a perfectly rectangular matrix for the dataset as there is for data sets of fixed format.

FREQUENCIES: an SPSS/PC+ procedure used to produce frequency counts for each value of each variable specified, together with percentages. Subcommands can also be used to request various formats for the output, including barcharts and histograms. Many optional statistics are also available, and these must be requested by name (see list in Chapter 6) rather than by number.

FREQUENCY CURVE: a plot of the distribution of values for a variable. A graphical representation of the frequency distribution.

FREQUENCY DISTRIBUTION: a set of scores in which data are grouped into classes and the frequency obtained for each class.

FRIEDMAN: an SPSS/PC+ subcommand (of the NPAR command) which requests a Friedman test. *See* Chapter 13.

FUNCTIONS: most commonly, transformations of numbers (e.g. square root, log). Functions can be used within SPSS/PC+ COMPUTE, IF, and SELECT IF commands. A large range of functions can be employed within SPSS/PC+, including numeric functions (e.g. ABS for absolute value, SQRT for square root), missing value functions (e.g. SYSMIS), etc. The full range of functions is given in the COMPUTE command section of the *SPSS/PC+ 4.0 Base Manual*.

FUNCTION KEYS: a set of special keys on the computer keyboard. When the SPSS/PC+ editor program REVIEW is in operation, these have special effects. These effects differ depending on whether the function key is pressed in isolation, whether the key press is followed by another key press, or whether the function key is pressed in association with another key such as the Alt. Throughout this book, an instruction to press a function key is indicated in the following way: **<F1>**. An instruction to press a function key followed by another key is shown as, for example: **<F9><w>**. And an instruction to press the function key in association with another key is shown as, for example: **<Alt-F10>**. A summary of the REVIEW commands involving the use of function keys will be found in *Appendix K*.

GE: a relational operator meaning 'greater than or equal to'. Thus IF (VAR1 GE 3) means 'if the value of VAR1 is greater than or equal to 3'.

GET: an SPSS/PC+ command which reads a system file previously written to disk with a SAVE command. If no filename is provided the file with the default name SPSS.SYS is read.

GLOSSARY: Apart from the glossary provided in the *SPSS/PC+ Base 4.0 Manual*, and the more extensive glossary provided in this book, SPSS/PC+ provides an on-line glossary of over 1200 terms which can be consulted only when using REVIEW. The command **<F1><g>** is used to switch the glossary function on. A record of any definition displayed can be 'pasted' into the file being edited. *See* Chapter 10.

GRAPHICS : an optional module of the SPSS/PC+ system which produces high quality pictorial representations of features of the data. Among the many available display formats are pie charts, bar charts, line graphs and scatterplots. GRAPHICS provides the user with an extensive range of options for controlling the size, shape and shading of graphs and, with a suitable printer, the system can produce colour graphics. In recent versions *SPSS/PC+ Graphics* includes integration with *Harvard Graphics*, a widely used and extremely flexible graphics program.

GT: a relational operator meaning 'greater than'. Thus IF (VAR1 GT 3) means 'if the value of VAR1 is greater than 3'.

HARD COPY: a printed copy of the contents of a file. It is useful to have a hard copy of all data files and definition files, and to keep hard copies of SPSS/PC+ output (e.g. printouts of edited listing files).

HARD DISK: technically, a Winchester disk unit. A rigid disk which stores information magnetically. Most hard disks store at least 40 megabytes of information – equivalent to that contained in 120 novels. The SPSS/PC+ system is stored on the hard disk.

HARDWARE: The physical components of the computer system, including the disk drives, keyboard, visual display unit, etc.

HELP: help with procedure commands and subcommand options, etc. can be gained when using REVIEW. The command **<F1><m>** switches the menu system on. The system then takes note of the context in which the request is being made (i.e. the command or subcommand currently being edited) and provides relevant information in a window at the top of the screen. One option will be highlighted, and by moving the highlight and selecting terms, particular options can be added to the command being written. The menu system is rather complex and is arranged in many different 'layers'. Switching the menu system off and returning to manual editing is achieved by pressing **<Alt-m>**. *See* Chapter 10.

HI (or **HIGHEST**): a keyword which can be used within a RECODE command, e.g.:

RECODE VAR1 (1 THRU 3 = 1) (4,5 = 2) (6 THRU HI = 3). HI should be used with care – if the highest value, or a value within a range specified by HI, is a user-missing value this too is recoded.

HISTOGRAM: an SPSS/PC+ subcommand (of the FREQUENCIES command) which requests that the frequencies of data be plotted as a histogram.

HISTOGRAM: a diagrammatic representation of data in which frequencies for different value-ranges of a variable are plotted graphically. Thus to plot a histogram of age, for a relatively small population with a large age range, we would first have to group data into value-ranges ('under 20', '20 to 30', etc.). Where there are only a few distinct values for a variable (for example, social class coded into 5 categories), such grouping of values is unnecessary and a bar chart may be used.

HYPOTHESIS: a tentative statement that describes a relationship between variables. An empirical study is needed to confirm or disconfirm the hypothesis. *See* null hypothesis.

IF: an SPSS/PC+ command which specifies that a transformation (the creation of a new variable or a change in values for an existing variable) should be made IF certain conditions are fulfilled. *See* Chapter 8.

IMPORT: an SPSS/PC+ command which reads a portable ASCII file. Portable files include both data and data dictionary information as well as certain system variables and system information. They can be transported from another computer using KERMIT and are then imported into the system when specified on an IMPORT command.

INCLUDE: an SPSS/PC+ operation command used to read a command file.

INDEPENDENT VARIABLE: a variable having two or more conditions in which a dependent variable is measured. Thus in an experiment, an independent variable (sometimes called the 'experimental variable' – say, the number of words in a list) is manipulated and a dependent variable (for example, the proportion of words remembered) is measured. The term is also applied in other types of study. Thus in an attitude survey the social class of the respondent may be treated as an independent variable. The attitudes examined in association with social class would be dependent variables.

INFERENTIAL STATISTICS: an area of statistical analysis which permits conclusions (inferences) to be drawn about whether a characteristic is significantly different in two or more populations, etc. More generally, the term applies to values calculated from a sample, and used to draw conclusions about the population from which the sample is drawn. This type of analysis contrasts with descriptive statistics which merely describe data (their frequencies, averages, 'spread', correlations, etc.). *See also* significance tests.

INITIALIZATION:

(a) (also referred to as 're-initialization') the process by which files with the names 'SPSS.LIS' and 'SPSS.LOG' are erased at the beginning of an SPSS/PC+ session. These are the default names for the listing file the log file. To keep a more permanent record of a session the user should provide different names for the listing and/or log files using the SET command at the beginning of a session, or should rename the log and listing files using the appropriate DOS command immediately after the SPSS/PC+ session has finished.

(b) the process of assigning a common value for a new variable to all cases before specifying (by means of a COMPUTE or an IF command) that certain cases should take specific values. *See* Chapter 8.

IN-LINE DATA: data that are included within a data definition file or a command file, or are entered interactively during a session. A BEGIN DATA. command must be placed immediately before the first line of data, and an END DATA. command placed on a line immediately after the last line of data. The in-line data method is not recommended. It is preferable for data to be entered into a separate data file and for this file to be read into the system when it is specified on a data list command included in a command file.

INPUT: to enter information into a computer.

INSTALLATION: the process of copying the diskettes on which SPSS/PC+ is originally supplied onto the hard disk and setting certain parameters of the computer system to make it suitable for SPSS/PC+ analysis.

INTEGER VARIABLE: a variable which can take only whole number values. Any variable which is a count is an integer variable. Thus the number of children in a family must be 1 or 2 or 3, etc. and cannot be 1.56 or 2.35.

INTERACTION: when the effects of an independent variable on a dependent variable are affected by the presence of another independent variable, the two independent variables are said to interact.

INTERACTION EFFECT: the extra combined effect of two or more independent variables on a dependent variable after the main effects have been taken into account. The term is used, for example, in two-way analysis of variance designs.

INTERACTIVE MODE: *see* interactive operation.

INTERACTIVE OPERATION: the issuing of commands directly to the machine in response to the SPSS/PC+ prompt. This mode of operation contrasts with the alternative mode (known as 'batch processing') in which a sequence of commands is written into a command file and this file then included.

INTER-QUARTILE RANGE: *see* quartile.

INTERVAL SCALE: a scale of measurement in which the data are truly quantitative (unlike nominal or ordinal data) and for which arithmetic operations such as addition, subtraction, multiplication and division are permissible. For an interval scale, equal *differences* between two numbers (e.g. the difference between 3 and 5, and between 10 and 12) signify equal differences in the quantity being measured. Unlike a ratio scale, however, the interval scale does not imply that there is a fixed zero-point. Parametric statistics can be used with interval data and ratio data. A general discussion about scales of measurement will be found at the beginning of Chapter 13.

JOIN: an SPSS/PC+ command which allows the user to combine two, three, four or five files (one of which can be the active file – the others must be system files) to become a new active file. JOIN MATCH effectively adds two (or more) sets of variables for the same (or overlapping) cases. JOIN ADD adds cases (with overlapping variables) to form a combined population. *See* Chapter 15.

K-S: an SPSS/PC+ subcommand (of the NPAR command) which requests a Kolmogorov-Smirnov test. *See* Chapter 13.

K-W: an SPSS/PC+ subcommand (of the NPAR command) which requests a Kruskal-Wallis test. *See* Chapter 13.

KEEP: an SPSS/PC+ subcommand used with the commands IMPORT and EXPORT, and JOIN. It instructs the system to include only specific variables when executing the operation. Thus if, with an original portable file of 100 variables, the IMPORT command includes a KEEP subcommand listing only 65 of these variables, the resulting active file will contain only those 65 variables. When KEEP is used, the variables in the new file are placed in the order in which they are specified by the KEEP subcommand. *See also* drop.

KENDALL: an SPSS/PC+ subcommand (of the NPAR command) which requests a test for Kendall's Coefficient of Concordance (*W*). *See* Chapter 13.

KERMIT: a file transfer program which allows SPSS portable files (including files originally created on the SPSS/PC+, SPSS/PC or SPSS^{X} systems) to be transferred between computers. Compatible versions of Kermit must be available on *both* of the machines. Kermit is an error-correcting program which monitors and checks the information-transfer process.

KEY DISKETTE : a special diskette (or 'floppy disk') which, in some circumstances of purchase, must be placed in Drive A in order for SPSS/PC+ to operate. The Key Diskette provides a way of copy-protecting the valuable system and thus prevents unauthorized use. Most large agencies and colleges which use SPSS/PC+ extensively will have purchased a site licence for the system, however, and the software provided will not use the key diskette protection method. Chapter 3 ('SPSS/PC+ and the Computer') makes the assumption that the user is *not* required to use a key diskette to operate the system.

KEYBOARD: the typewriter-like part of the computer system which allows the user to input information. It includes the alphabetic keys and a number of additional keys – including a number pad, the control key (**<Ctrl>**, the Alt key (**<Alt>**), an escape key (**<Esc>**), and a number of function keys (**<F1>**, **<F2>**, etc.).

KEYWORD: a word which has a special assigned meaning in SPSS/PC+. As well as command and subcommand words, certain other words, including BY, WITH, TO, ALL, AND, THRU, ELSE and NOT, are keywords. It is important to avoid keywords when naming variables.

KURTOSIS: the extent to which the data peaks around the middle value or is flat across the range of frequencies. The standard error of the kurtosis is a measure of the confidence we can have that the kurtosis of the current sample is a true representation of the kurtosis of a larger theoretical population.

LABEL: an optional string of characters used to provide annotations for SPSS/PC+ output. Variable labels (specified on a VARIABLE LABELS command) are used to provide an extended description of variables. Value labels (specified on a VALUE LABELS command) allow the user to provide brief descriptions of the particular values which a variable may take. Labels must always be enclosed in apostrophes (or quotation marks) when specified in commands. Although labels may be up to 60 characters in length it is best to keep them as succinct as possible. All labels are optional. If a spelling error is included in a label specification it will have no effect on the operation of the program (but, of course, the labelling within the output will include the erroneous spelling).

LE: a relational operator meaning 'less than or equal to'. Thus IF (VAR1 LE 3) means 'if the value of VAR1 is less than or equal to 3'.

LEAST SQUARES METHOD: a method of determining the form of curve (of a specified kind) which best fits a set of data.

LEVELS OF MEASUREMENT: *See* scales of measurement.

LIKERT SCALE: usually, a three- or five-point scale in which the respondent indicates the degree to which s/he agrees with a particular statement. The points on the scale are often labelled 'Strongly agree', 'agree', etc.

LINEARITY: the degree to which the relationship between paired scores on two variables can be represented graphically by a straight line.

LINEAR REGRESSION: a procedure for finding a simple formula to predict values of a dependent variable from values of an independent variable. By minimizing errors of estimate, the best-fitting (least squares) straight regression line is found. If a straight line is not a good representation of the relationship, linearity may be achieved by transforming data. Values for the dependent variable can be predicted from scores on the independent variable by an equation which represents the slope of the line and the point at which it intercepts an axis on the graph. In some cases a relationship is best represented not by a straight line but by a curved line (a 'curvilinear relationship') such as a parabola. The REGRESSION procedure (part of the *SPSS/PC+ Statistics 4.0* module) can be used to assess linearity, to find the best fitting regression line and to predict values for the dependent variable. *See also* multiple regression.

LISTING FILE: a text file which is automatically written by SPSS/PC+ and contains any output which has appeared on the screen (or in printed form) as a result of procedures activated during the session (including commands issued, warnings and error messages, and also page numbers and titles). The default name for the listing file is SPSS.LIS, although this can be changed by using an appropriate SET command. Any file named SPSS.LIS within the current computer directory is initialized when the next SPSS/PC+ session begins. Listing files are stored in ASCII form and can be printed or edited. It is useful to rename the listing file immediately after the end of an SPSS/PC+ session, using the DOS command for renaming a file, and to make a copy to a diskette for editing. This prevents the loss of valuable output information when initialization takes place at the beginning of the next session.

LISTWISE DELETION: an SPSS/PC+ term describing one way in which missing data are excluded from analyses. If a particular case has missing data for any variable included in a list of variables, then the case is excluded from all procedures and calculations specified for that list. This strategy contrasts with that of pairwise deletion.

LO (or **LOWEST**): a keyword which can be used within a RECODE command, e.g.: RECODE VAR1 (LO THRU 2 = 1) (3,4,5 = 2) (6,7 = 3). LO should be used with care – if the lowest value, or a value within a range specified by LO, is a user-missing value or the system-missing value this too is recoded.

LOCATION: the position within the dataset of the data for a particular variable. In fixed format, data for a particular variable must be located in the same column(s) for all cases.

LOG FILE: a text file which is automatically written by SPSS/PC+ and contains any commands entered during the SPSS/PC+ session. The log file also contains information that warnings or error messages were issued, and provides references to output page numbers. A log file may be edited within an SPSS/PC+ session, using REVIEW. Thus a series of commands issued interactively can be modified and then written to disk as a command file. The default name for the log file is SPSS.LOG, although this can be changed by the use of an appropriate SET command. Any file in the current directory with the name SPSS.LIS is initialized (overwritten) when the next SPSS/PC+ session begins. Log files are ASCII files.

LOGICAL EXPRESSION: an expression which is evaluated for its truth value. It is frequently encountered in a conditional transformation. Thus: IF (A = B) C = 1. Here 'A = B' is the logical expression. If, for a particular case, the expression is TRUE (i.e. the value of A *does* equal the value of B), then, for that case, C will be set to equal 1. Logical expressions may be very complex and may include many relational operators (equals, less than, etc.). Two or more expressions may be joined by one or more logical operators.

LOGICAL OPERATOR: an operator used in conditional transformations to join two or more logical expressions. SPSS/PC+ employs the logical operators NOT, AND and OR. *See* Chapter 8.

LONG STRING: a string variable from 9 to 255 characters long. Each segment of 8 characters counts as one variable towards the system limit of 500 variables. *See also* string variable; short string.

LT: a relational operator meaning 'less than'. Thus IF (VAR1 LT 3) means 'if the value of var1 is less than 3'. The 'less than' symbol (<) can be used to replace LT.

MAIN EFFECT: the effect of a single independent variable on a dependent variable (for example, in an analysis of variance). *See also* interaction effect.

MAP: an SPSS/PC+ subcommand used within the commands IMPORT and EXPORT, and JOIN. It produces a listing of the dictionary of the current active file.

When used with a JOIN command, MAP also specifies the original files from which each of the variables in the composite file has come. With EXPORT and IMPORT, MAP lists the actions implemented up to that point (thus allowing the user to keep track of variables which have been referred to in DROP, KEEP or RENAME subcommands). MAP can be used several times within a single command; if it is placed as the last subcommand then it provides a list of the final contents of the active file.

MATRIX: a two-dimensional table of data values or results. For some SPSS/PC+ procedures (e.g. CORRELATION, ONEWAY) matrices may be written to a results file by specifying a particular OPTION. Such a matrix may then be used as input to another procedure, increasing the efficiency and speed of the analysis. Data that are entered into a procedure in this way are referred to as 'matrix data'.

MATRIX DATA: some SPSS/PC+ procedures (e.g. ONEWAY and REGRESSION) will accept summary statistics (i.e. the results of previous analyses of the relevant data) as their input data. If data is to be entered in this form a MATRIX specification must be used on the DATA LIST command.

MAXIMUM: the highest value for a particular variable within a set of data.

MCNEMAR: an SPSS/PC+ subcommand (of the NPAR command) which requests a McNemar test. *See* Chapter 13.

MEAN: the arithmetic average. The sum of all scores divided by their number. The standard error of the mean is a measure of the confidence we can have that the average based on the present sample is a true representation of the average of a larger theoretical population.

MEDIAN: an SPSS/PC+ subcommand (of the NPAR command) which requests a Median test. *See* Chapter 13.

MEDIAN: the midscore. The value which is a mid-point, dividing the population into 50% below the median and 50% above the median.

MINIMUM: the lowest value for a particular variable within a set of data.

MISSING DATA: if, for a particular case, the value for a variable is unknown (for example, if a person has not answered a question on a questionnaire) then the relevant data are 'missing.' Such circumstances are represented in the dataset by a missing value which has been assigned for that variable by the user. This is the 'user-missing value'. In certain circumstances a missing value is assigned to a variable by the system. For example, when a new variable is created by a COMPUTE command, any case for which the variable cannot be computed is assigned a missing value for that variable. This is the system-missing value, and it is indicated in screen displays and printed output by a period.

MISSING VALUE: an SPSS/PC+ data definition command that provides a value (or code) which, when encountered in the data for that variable, is interpreted by the system as indicating that the real value (for this variable and for this case) is unknown (i.e. 'missing'). Data having the missing value are ignored during statistical calculations (unless the system is instructed otherwise).

MODE: the most frequent value for a variable.

MODULE: the unit for the removal and installation of SPSS/PC+ procedures. The various SPSS/PC+ procedures (both those included in the base system and those included in the optional modules) are grouped together into modules. Some of these are non-permanent and can be removed from the system and re-installed using the SPSS MANAGER procedure. *See* SPSS MANAGER.

MONITOR: the visual display unit. The screen displays both the input to the system and the output from the system. SPSS/PC+ can also send output directly to a printer. Screen displays can be suppressed by using a SET command.

MORE: in many cases it is not possible for all of the information provided by SPSS/PC+ (results, help information, etc.) to be accommodated on a single screen. The first screenful of information is displayed and the operation then paused to allow the user time to read the contents. At the same time the message 'MORE' is displayed. When any key is pressed, the next screen of information is displayed. If the results of a set of procedures will extend over many displays, the user may wish to read the results from printed output, rather than from the screen. In this case the MORE function would lead to repetitive key pressing and would serve no useful purpose. The MORE operation may be switched off by entering the SET command 'set more off.'.

MOSES: an SPSS/PC+ subcommand (of the NPAR command) which requests a Moses test. *See* Chapter 13.

MS-DOS: the disk operating system (*see* DOS) used by SPSS/PC+. This set of programs controls the interaction between different hardware elements within the system, including the operation of peripherals (printer, screen, disk drives, etc.), and controls such functions as the copying, deleting, and printing of files.

MULTIPLE COMPARISON TESTS: statistical tests used (for example, after a one-way analysis of variance – using the SPSS/PC+ procedure ONEWAY) to determine which pairs of groups are significantly different. Thus if an analysis of variance has shown that there are overall differences in height between five samples from different ethnic groups, multiple comparison tests will show which particular groups differ significantly from which others. The multiple comparison tests performed after a general effect has been identified are sometimes referred to as '*post-hoc* tests'. Multiple-comparison procedures impose stringent criteria for significance, to correct for the fact that the probability of finding 'significant' differences that are actually due to chance alone increases with the number of comparisons made. In the ONEWAY procedure seven multiple comparison tests are available (Duncan's multiple range test, the Tukey test, etc.). Each must be requested using a separate Ranges subcommand.

MULTIPLE RANGE TESTS: *See* MULTIPLE COMPARISON TESTS.

MULTIPLE REGRESSION: a statistical technique for estimating simultaneous correlations between multiple predictor variables and a single dependent variable. Multiple regression may be requested by the REGRESSION procedure which forms part of the *SPSS/PC+ Statistics 4.0* module.

MULTIVARIATE ANALYSIS: the name given to any statistical technique which attempts to identify the component structure of complex data. Examples of such analyses include factor analysis, cluster analysis, log-linear analysis and multivariate analysis of variance (all of which are included in either the *SPSS/PC+ Statistics 4.0* or *SPSS/PC+ Advanced Statistics 4.0* modules).

M-W: an SPSS/PC+ subcommand (of the NPAR command) which requests a Mann-Whitney U test. *See* Chapter 13.

N: an SPSS/PC+ selection command which places a limit on the size of the active file to the first N cases. The command takes the form 'N 54'. *See* Chapter 8 for further information.

NE: a relational operator meaning 'not equal to'. Thus: IF (VAR1 NE 3) means 'if the value of VAR1 is not equal to 3'.

NOMINAL SCALE: a scale of measurement in which data are grouped into 'name' classes between which there are no formal mathematical relationships. Thus although classes are different from each other there is no implied order of the categories. Although categories may be assigned numerical codings (e.g 1 for male and 2 for female) such numbers act only as labels and should not be used in calculation. Certain non-parametric statistics including the chi-square test, the binomial test and the Cochran test, are suitable for the analysis of nominal data.

NON-PARAMETRIC STATISTICS: statistical methods which do not assume that data are distributed normally. Frequently used non-parametric methods include the chi-square test and other tests included in the SPSS/PC+ NPAR procedure. Rank-order correlations can be obtained by using the RANK procedure followed by CORRELATION. *See* normal distribution; Chapter 13.

NORMAL CURVE: also known as the 'Gaussian curve' and the 'probability curve.' A plot of the normal frequency distribution. *See* normal distribution.

NORMAL DISTRIBUTION: a frequency distribution which conforms to a standard 'bell-shape'. The three measures of central tendency – the mode, mean and median – are at the same point, and the distribution is symmetrical. If a set of data conforms (or nearly conforms) to this distribution then statistical inferences can be made about it using parametric statistics.

NOT: a logical operator used, usually in a conditional transformation, to reverse the true-false status of the logical expression which follows it. Thus 'IF (NOT VAR1 EQ 2) GROUP = 1'.

NPAR: an SPSS/PC+ procedure for the non-parametric analysis of data. NPAR includes a number of different tests for the analysis of categorical and ordinal data from one or more groups. *See* Chapter 13.

NULL HYPOTHESIS: the hypothesis that any statistical difference or association obtained will be the result of mere chance. If the result of an analysis falls outside the range which can reasonably be attributed to chance fluctuations then that result is said to be 'statistically significant' and the null hypothesis is therefore rejected. The conclusion in such a study would be that the effects are *not* merely the result of chance. The indirect strategy of rejecting the null hypothesis is linked to assumptions underlying statistical procedures. Sometimes abbreviated to H_o.

NUMERIC VARIABLE: a variable for which the values are represented by true numbers (rather than codes). Data for such variables can therefore be used legitimately in arithmetic calculations.

NUMERICAL EXPRESSION: an expression which consists of numbers or numeric variables linked by operators such as +, – and /. Thus '(VAR001 + 6 – VAR005)' is a numerical expression. Such an expression might be used, for example, in creating a new variable with a COMPUTE command.

ONE-TAIL TEST: a test for the significance of a difference in a case in which the hypothesis has been cast in a directional form. Contrasts with the two-tail test of a non-directional hypothesis. Using a one-tailed test, the null hypothesis is rejected if there is a small probability (the criterion is often .05 or less) of obtaining a result as extreme as the one observed, and in the same direction, when the null hypothesis is true. *See* null hypothesis; tail; significance tests.

ONEWAY: an SPSS/PC+ statistical procedure used to perform a one-way analysis of variance.

ON-LINE: 'immediately accessible'. It is usual for computer programs to run one at a time, but sometimes one program, or a different part of the same program, may be called into use while the other is still active. Where such a situation occurs the second program or the second part of the same program is said to be available 'on-line' – it is immediately accessible and will not interfere with the program currently in operation. Thus SPSS/PC+ allows the user, when editing a file with REVIEW, to access on-line help (using the command **<F1><m>** to switch on the menu system) or the on-line glossary (using **<F1><g>**).

OPEN-ENDED QUESTION: a question which permits the respondent to answer in free form, not limiting the response to a choice between given alternatives. Examples: 'Where were you born?'; 'Name three things you like' and 'Give any ideas you have for improving health care.' Such questions often provide rich information, but the responses may be difficult to analyse. A coding system may be devised to classify or score the responses. Open-ended questions contrast with closed-ended questions.

OPERATOR: a symbol denoting a mathematical or logical process. Operators are used in various SPSS/PC+ commands, and especially in conditional transformations and selections. We can distinguish between arithmetic operators (+, /, etc.), relational operators (EQ, NE, GT, etc.) and logical operators (AND, OR and NOT).

OPTIONS: a subcommand available with many procedures which allows the user to specify various features of the analysis and output. The OPTIONS subcommand often allows the user to specify how missing data should be treated, the form the analysis should take, and how the output should be formatted and labelled. For some procedures, the OPTIONS subcommand is also used to specify that input data should be read from a matrix, or that results should be written to a results file in matrix form.

OR: a logical operator used to join two or more relations in a conditional transformation. Thus 'IF (VAR1 EQ 2 OR VAR1 EQ 3) ...' and 'SELECT IF (VAR1 EQ 2 OR VAR1 EQ 3).' Notice that such a statement is written in a 'full' form . The short form: 'IF (VAR1 EQ 2 OR 3)' would *not* be valid.

ORDINAL SCALE: a scale of measurement in which the numbers refer to positions in an ordered series. Thus in a race, the results in terms of first, second, third, etc. may be represented by 1, 2, 3, etc. but these are ranks and should not be regarded as real numbers which can be added, subtracted, multiplied, etc. If two or more items share the same placing then there are said to be 'tied ranks'. Certain non-parametric statistics, including the Moses test and the Wilcoxon test are suitable for use with ordinal data and may be requested within an NPAR command. Rank-order correlations can be obtained by using the RANK procedure followed by CORRELATION. *See* scales of measurement; Chapter 13.

ORTHOGONAL: 'forming right angles'. If there is no association between two variables or two factors resulting from a factor analysis (i.e. if the correlation between them is 0) they are described as orthogonal.

OUTPUT FILE: a file containing information formatted in a specific way so that it can be read by a computer program as 'meaningful' data. Thus some SPSS/PC+ procedures create output files containing a data matrix formatted in a specific way. SAVE produces a system file that can later be read with a GET command, and EXPORT produces a portable file that can be read by a range of SPSS programs. The term is also loosely applied to the listing file automatically produced by SPSS/PC+ for every session. The listing file includes all of the output (displays of tables, statistical analyses, etc.) produced during the session (*See* listing file).

PAIRWISE DELETION: an SPSS/PC+ term describing one way in which missing data are excluded from analyses. If a particular case has missing data for one or both of two variables specified for a particular procedure that case is excluded from the calculation. Otherwise it is included. This strategy contrasts with that of listwise deletion.

PARALLEL FORMS RELIABILITY: *See* alternate forms reliability.

PARAMETER: any measurable characteristic of a population.

PARAMETRIC STATISTICS: statistical methods which make certain assumptions about the parameters of the population from which the sample has been drawn. In particular, they assume that observations are independent and that measurements are continuous. They may also assume that the data frequencies are normally distributed (or approximate to such a distribution). Statistics such as the *t*-test, product-moment correlation and analysis of variance depend on the data having such characteristics, although in practice such tests are able to return meaningful results even with data for which certain of the assumptions do not hold.

PARTIAL CORRELATION: the correlation between two variables which remains when the effects of certain other variables have been 'partialled out' (i.e. eliminated or corrected for).

PATH COMMAND: A command that specifies a route through the computer directory hierarchy to a subdirectory or file. Once a path command is in operation, files in the specified subdirectories can be accessed from any directory. Thus if a path command specifying the SPSS directory is in operation, SPSS/PC+ will be accessible from any directory (e.g. the user's personal directory) simply by entering the command 'spsspc'.

PEAK: the highest point in a frequency distribution.

PERCENTILE: a range which contains a hundredth part of an ordered dataset. If the data has been arranged in a frequency distribution of ascending values then the value which cuts off the lowest valued hundredth part of the set is the 'first percentile'. The 'fiftieth percentile' will correspond to the median. If the sample is normally distributed then percentile values can be calculated from a knowledge of the mean and the standard deviation.

PLOT: an SPSS/PC+ procedure which produces a graph of the frequencies of data for two variables simultaneously. *See* Chapter 11.

POISSON DISTRIBUTION: a highly asymmetric frequency distribution which is obtained when very rare random events are plotted.

POPULATION: a complete set of objects, individuals or measurements which have some characteristic in common. A group from which a sample may be drawn.

PORTABLE FILE: an ASCII file produced by SPSS/PC+, SPSS/PC or $SPSS^X$, using the EXPORT command. Portable files include a copy of the data, data definition information, system variables and system information. The file transfer program Kermit is used to transfer portable files between mainframe computers running $SPSS^X$ and microcomputers running SPSS/PC+. Portable files are not easily editable. *See* Chapter 16.

PROCEDURE: a type of SPSS/PC+ command, often a request for a statistical analysis, which reads data values from the active file. If no active file is current, the procedure command will create such a file by following any 'data read' commands which have previously been issued. Before beginning the requested analysis, procedure commands execute any transformations which have been requested but not yet put into effect.

PROCESS IF: a command which selects cases to be included in the next procedure if certain conditions are fulfilled. The conditions are specified by a logical expression which may include relational operators (such as NE – not equal to, or LT – less than) and logical operators. Thus: PROCESS IF (VAR1 NE 5 OR VAR2 LT 7). PROCESS IF leads to a selection for the *next procedure only.* It does not lead to cases being discarded from the active file, whereas the command SELECT IF permanently reduces the number of cases in the active file. *See also*: selection.

PRODUCT-MOMENT CORRELATION: a test designed by Karl Pearson to determine whether high values of one variable are associated with high scores on another (and low values on one with low values on the other). The calculation produces the statistic *r*, the 'correlation coefficient', which varies between -1 (a perfect negative correlation) to +1 (a perfect positive correlation). This statistic is calculated by the SPSS/PC+ CORRELATION procedure.

PROFILE: a specific tailoring of the SPSS/PC+ environment by a series of commands (usually SET commands). If the user prefers to work with an environment different to that set by default, for example with MORE off rather than on, or with different colours on the screen, or with the listing file named something other than SPSS.LIS, then s/he may write a batch file which includes the relevant commands. This may be brought into action with an INCLUDE command at the beginning of a session. However, if such a file of commands is saved to the disk with the name "SPSSPROF.INI" it will be automatically executed each time that SPSS/PC+ is run from the current directory. The user who works from his or her own directory is thus able to create an automatic profile by writing into the SPSSPROF.INI file a series of SET commands which reflect personal preferences in working with the system. SPSSPROF.INI is read from the directory from which the user enters SPSS/PC+. The commands are displayed immediately after the SPSS/PC+ logo. Different users may have quite different SPSSPROF.INI files in their own directories.

PROMPT: an instruction which appears on the screen to give information and invite a response from the user. A number of common screen prompts are used by SPSS/PC+. When the system is 'at rest' waiting for a command, the SPSS/PC+ prompt 'SPSS/PC:' is displayed. Several prompts are given in response to REVIEW commands. Help with responding to such a prompt can be obtained at this point by pressing **<F1>**.

QUARTILE: one of three values which divide a frequency distribution into equal quarters. These values will depend on the range and the shape of the frequency distribution. The first quartile marks the '25th percentile', the second marks the '50th percentile' and the third marks the '75th percentile'. The distance between the first and third quartiles (therefore spanning the range containing the central 50% of the distribution) is known as the 'inter-quartile range'.

QUOTA SAMPLING: a sampling method by which respondents are selected to be representative of a larger population with respect to certain (usually demographic) characteristics such as age, social class and sex. The researcher decides how many of each category of case should be sampled and then samples cases in each category until the specified quota is filled. This method is mostly used in market research.

RANDOM: occurring by chance alone.

RANDOM ERROR: anything that increases the variability of the measurement of a dependent variable and does not change systematically with different levels of an independent variable. Such error is found to be normally distributed around the mean.

RANDOM SAMPLE: a selection chosen at random from a population and assumed to be representative of it.

RANGE: the difference between the lowest value and the highest value for a variable.

RANGES: a subcommand used (within the ONEWAY procedure) to request one of seven statistical 'multiple comparison tests'. The request for each of these tests must be specified on a separate RANGES subcommand.

RANK: the position of an item in an ordered series.

RANK: a procedure which ranks any specified variable and creates a new variable in the active file, the values of which correspond to either an ascending rank-ordering or a descending rank-ordering of the original variable. By default, the name of the variable created is the original variable name prefaced with an r, and the new variable is given the label "RANK of (ORIGINAL)". RANK can also be used to order a particular variable within specific groups that make up the population. Various strategies are available for dealing with tied ranks. *See* Chapter 14.

RANK-ORDER: To assign ordinal positions to cases on the basis of their values on a particular variable. *See* ascending ranking; descending ranking; tied ranks.

RANK-ORDER CORRELATION: A correlation between pairs of ordinal (ranked) variables. Using SPSS/PC+, a rank-order correlation (for example the Spearman rank correlation coefficient) may be obtained in the following way. First the RANK procedure is used to create new ordinal variables from existing variables, and the CORRELATION procedure is then used to obtain correlations between pairs of the newly-created ordinal variables. Care must be taken to specify either an ascending rank-ordering or a descending rank-ordering in the creation of the new variables. This critically affects the nature of the created variables and the interpretation of a positive or negative correlation coefficient.

RATING SCALE: a means by which a respondent can record his or her judgement according to a numerical or 'analogue' system with well-defined end-points. Thus a scale for judging pain might involve the end-point descriptions 'no pain' and 'intense pain'. Gradings might be made on such a scale in terms of a fixed number of points (7-point scale, 11-point scale, etc.). Alternatively, the respondent might be asked to mark an appropriate point on a continuous line to indicate the degree of pain involved. The response would then be scored by measuring the distance of the respondent's mark from the end of the line.

RATIO SCALE: a scale of measurement in which data are truly quantitative. Ratio data share with interval data the fact that equal *differences* between numbers (e.g. the difference between 3 and 5, and between 10 and 12) are equal. For ratio data, however, the zero point is not arbitrary but represents the true absence of the quality being measured. Height in centimetres, for example, is a ratio measurement. A ratio scale represents the highest level of measurement.

RAW DATA: the data as they are originally obtained, i.e. before they have been subjected to any transformation, coding or grouping.

RECODE: an SPSS/PC+ command which is used to revise the values of an existing variable. Thus if the user needs to compare an 'old' group with a 'young' group, and *age* is currently coded in terms of years, the following RECODE could be used to produce just two age-groups: RECODE age (10 thru 30 = 1) (31 thru 98 = 2). *See* Chapter 8.

REGRESSION: linear regression is a relationship between two variables when their values, plotted graphically, produce a straight line. Multiple regression is the relationship between a single 'criterion variable' and a number of 'predictor variables.'

REGRESSION: an SPSS/PC+ procedure for performing linear and multiple regression. It is included within the *SPSS/PC+ Statistics 4.0* module (*see* Chapter 16 for a brief description).

REGRESSION ANALYSIS: An analysis that provides an estimate of the relationship between a dependent variable and one or more independent variables or covariates.

RELATIONAL OPERATOR: a symbol that is used in a logical expression to compare one value with another. Thus one value may be equal to (EQ) another, not equal to (NE), greater than (GT), or less than (LT) another, etc. Relational operators are used in SPSS/PC+ conditional transformations (IF, SELECT IF and PROCESS IF). *See* Chapter 8.

RELIABILITY: the consistency of a test (in terms of internal consistency and stability). The degree to which a score on a test reflects the 'true' score rather than error. Thus reliability can be regarded as the ratio of true variance to observed variance. Reliability may be assessed in a variety of ways, including the test-retest method, the alternate forms method, the split-half method and the internal consistency method. SPSS/PC+ provides a procedure, RELIABILITY, which allows reliability of multiple-item additive scales to be assessed. *See also* alternate forms reliability; alpha; internal consistency; split-half reliability; test-retest reliability. *See* Chapter 14.

RELIABILITY A procedure used to estimate the reliability of a scale formed by adding scores for a number of variables within the active file. Any of five models may be used to assess reliability, including the internal consistency model and the split-half model. *See also* split-half reliability; internal consistency; alpha coefficient; equal length Spearman-Brown coefficient; unequal length Spearman-Brown coefficient; Guttman split-half coefficient *See* Chapter 14.

RELIABILITY COEFFICIENTS: measures of the reliability of a scale (i.e. how far the score obtained for a case reflects the true score for the case).

REPEATED-MEASURES: a research design that involves the same variable being measured for the same case on several different occasions is following a repeated measures design. The advantage of such a design is that error variance due to differences between subjects is eliminated. However, there may be certain drawbacks in using such a design (for example, an effect observed on one measurement occasion may be due to 'latent effects' left over from an experimental manipulation instigated before a previous measurement occasion). Nevertheless, in many cases a repeated measures design is very powerful. The paired *t*-test provides one simple example of a repeated measures design. To combine the power of analysis of variance with the power of repeated measures, however, the multivariate analysis of variance procedure MANOVA must be used. The MANOVA procedure is part of the *SPSS/PC+ Advanced Statistics 4.0* module.

REPORT: an SPSS/PC+ procedure used to produce tables, including listings of cases and summary statistics. Before a REPORT command is issued the data must be organized into an appropriate order using a SORT command. Data can be broken down by 'break' variables (using a BREAK subcommand) and the display is then specified by one or more SUMMARY subcommands. Various labels may also be specified. The use of the REPORT procedure is not covered in this book.

RESPONDENT: usually, a person who completes or answers a questionnaire. More generally, any case in the sample that has been surveyed or observed.

RESPONSE RATE: the number of returned questionnaires as a proportion of those sent out.

RESULTS FILE: a file which includes results from a specific procedure. It can be produced using the WRITE command or by specifying an appropriate OPTION with certain procedures (for example CORRELATION and ONEWAY). By default, the results file is named 'SPSS.PRC'. This can be changed using the SET command. Although results files *can* be edited with an editor any changes made in this way may make them unsuitable as input to other procedures. Editing of these files is therefore generally not recommended. If the user needs an editable file which includes results of procedures, the listing file will generally prove more suitable.

REVIEW: the SPSS/PC+ text editor which allows the editing of text files (including data files, data definition files, command files, log files and listing files). REVIEW can only be used for editing ASCII files. REVIEW can be entered directly from DOS (*see* Chapter 4 for details) or at the SPSS/PC: prompt: e.g. REVIEW "gender.dat". Within REVIEW it is possible to copy, move, and delete lines and areas (blocks of whole lines or rectangles). It is also possible to find and change strings. Whole or part of the edited file can be saved, and other files can be entered into the file being edited. Editing and other functions are largely controlled by the use of function keys, which sometimes need to be pressed before or in association with another key. Thus **<Alt-F10>** is the command for exiting from REVIEW. The Alt key is held down while the F10 key is pressed. **<F9><w>** is the command for writing the whole of the file being edited to the disk. An on-screen guide to the various REVIEW functions can be obtained from within REVIEW by pressing **<F1>**. Also available from within REVIEW are the on-line help facility (**<F1><m>**) and the on-line glossary (**<F1><g>**). Pressing the relevant function key produces a minimenu at the bottom of the screen. A particular function may be requested by highlighting the appropriate term and pressing <RETURN>.

RUNS: a subcommmand requesting a runs test, a non-parametric test available within the NPAR procedure. *See* Chapter 13.

SAMPLE: a selection of scores from a larger (often 'theoretical') population. If the sample is random then it is assumed to be representative of the larger population.

SAMPLE: an SPSS/PC+ command which draws a temporary sample of cases from those currently in the active file. The usual form is: SAMPLE .33. This would lead to approximately one third of cases being selected.

SAMPLING ERROR: the variation in scores or statistics which derives from the non-representativeness of the sample.

SAMPLING PROCEDURES: strategies for selecting a sample effectively and efficiently so that valid conclusions can be drawn regarding a larger population. Various sampling methods are used. *See* area sampling; quota sampling; stratified sampling.

SAVE: a command which produces an SPSS/PC+ system file. The default name for a file created with SAVE is 'SPSS.SYS' but an alternative name can be specified (e.g. SAVE OUTFILE= "gender.sys"). There is also an opportunity, when saving a system file, to DROP some of the variables included in the current active file.

SCALE: an instrument used to assign scores, measurements of a particular construct or dimension, to cases (respondents). There are single rating scales and multiple-item scales. The reliability, internal consistency, and other characteristics of additive multiple-item scales can be assessed with the RELIABILITY procedure. *See* Chapter 14.

SCALES OF MEASUREMENT: also referred to as 'levels of measurement'. Numbers are used in different ways, and have different properties, depending on how they are used. Some numbers merely act as labels, some refer to ranks, and some represent true measures. A number which is a label, for example, cannot be added – it would be foolish to think that two football players numbered 1 and 3 were somehow equivalent to one player labelled 4. Different types of statistical analysis are appropriate when numbers are used in these different ways. 'Scales of measurement' refers to these different ways of using numbers. The most commonly distinguished scales are nominal, ordinal, interval, and ratio.

SCATTERPLOT: a graph in which items are plotted according to their values on two variables, one of which is scaled on the horizontal axis and the other of which is scaled on the vertical axis.

SCROLLING: the process of moving information up and down the screen in order to get to the part that needs to be examined or worked on.

SELECT IF: a command which selects cases to remain in the active file if certain conditions are fulfilled. The conditions are specified by a logical expression which may include relational operators (such as EQ – equal to, or GT – greater than) and logical operators. Thus: SELECT IF (VAR1 GT 3 OR VAR2 EQ 8). SELECT IF permanently affects the active file whereas the command PROCESS IF specifies which cases should be included in the next procedure (only). *See also*: selection.

SELECTION: the process of including only certain cases for analysis or manipulation (transformation, etc.). The two SPSS/PC+ commands SELECT IF and PROCESS IF may be used to select subsets of the original population according to criteria specified by the user. PROCESS IF is a temporary selection command and leads to a selection for the *next procedure only*. SELECT IF, however, makes a 'permanent' selection of cases. For example: SELECT IF (SEX = 1). where male cases were coded 1 on the variable *sex* would lead to only the male cases remaining in the active file.

SESSION: an SPSS/PC+ session begins when the 'spsspc' command is issued and the SPSS/PC+ logo is displayed on the screen, and it lasts until the command FINISH is issued in response to the SPSS/PC+ prompt. Within a session many different active files can be created, and many different analyses performed. If an active file is tailored (for example with a SELECT IF command, or a RECODE command) then the changes remain in effect for the rest of the session (unless the original file is made active again). Log files and listing files are maintained for each session and are over-written at the beginning of the next session (unless their names have been changed from the default names).

SET: an SPSS/PC+ command which allows the user to specify a number of options affecting the way in which the system behaves, e.g. the names to be given to the listing file and log file and the 'on/off' condition of the printer, the MORE feature and the BEEP. Thus to switch the printer on, the command would be: SET printer = on.

SHORT STRING: a string variable up to 8 characters in length. Such strings can be used in transformation commands and SORT commands but are not available for arithmetic operations. *See also* string variable; long string.

SHOW: an SPSS/PC+ command which produces a screen display of information about the current state of certain parameters concerning the running of the system. These include the memory space available, the current names for the log file and listing file, etc. Many of these parameters can be changed by an appropriate use of the SET command.

SIGNIFICANCE TESTS: some statistical procedures merely describe or summarize data – these are descriptive statistics. Others examine the significance of a difference or an association between samples – these are 'inferential statistics'. When a difference or association has been found we need to establish whether the magnitude of the effect is within the range which could reasonably be explained by mere chance factors or whether it is 'significant' i.e. unlikely to be a chance effect. In making an inference about a difference between two samples, for example, we are assessing the probability that both samples might have been drawn from the same population. One common criterion for accepting an effect as significant is that an effect (e.g. difference) of such a magnitude would occur by chance less than 1 in 20 times. If the result of the statistical procedure reaches this level it is said to be significant at the '5% level.' A result which would occur less than 1 in 100 times by chance is significant at the '1% level' (etc.). Traditionally the researcher, having calculated a value for the statistic (say t or chi-square) would consult a set of statistical tables to determine whether the value was significant. SPSS/PC+ automatically provides information about the significance of many of the statistics calculated. A statistically significant result may or may not be significant in other ways, and the validity of a statistical inference will always depend of the appropriateness of the test which has been used. It is possible, particularly with packages such as SPSS/PC+, to obtain complete statistical analyses – including estimates of statistical significance – on data which completely violate the assumptions which must be fulfilled in order for the test to be validly applied.

SIGN (TEST): a non-parametric test, available within the NPAR procedure, used to determine whether there is a significant difference between matched pairs of data. The test takes into consideration only the direction of the differences between pairs of data and ignores the magnitude of such differences. *See* Chapter 13.

SIGNED-RANKS TEST: a non-parametric test, also known as the Wilcoxon, available within the NPAR procedure. *See* Chapter 13.

SKEWNESS: a measure of the 'lop-sidedness' of a frequency distribution – i.e. a measure of the tendency for data to cluster towards the lower or upper end of the range rather than being evenly distributed or balanced around the mid-point. The 'standard error of the skewness' is a measure of the confidence we can have that the skewness of the particular sample is a true representation of the skewness of a larger 'theoretical' population.

SOFTWARE: computer programs. The SPSS/PC+ system is a highly complex and sophisticated program – software – run on the computer machinery – the hardware.

SORT: a procedure command which re-arranges cases into a particular order within the active file according to one or more 'key' variables (such as *age* or *surname*). The SORT command allows cases to be sorted alphabetically or numerically and in either ascending or descending order. 'Ascending', here, means 'low to high' (or, for an alphabetic sort, 'A to Z').

SPLIT-HALF RELIABILITY: reliability assessed by splitting a test into two halves and correlating the scores obtained from the two halves. This form of reliability can be requested within a RELIABILITY command. *See* Chapter 14.

SPSS.LIS: the default name for the listing file. This file is initialized at the beginning of each SPSS/PC+ session. The name of the listing file can be changed at any time within a session by issuing a suitable SET command. The listing from that point to the end of the session will then be stored on a file with the new name, and this will NOT be inititialized at the beginning of the next session. An alternative strategy for keeping listing files is to rename SPSS.LIS immediately after the end of a session, perhaps making a copy on a diskette.

SPSS.LOG: the default name for a log file. This file is initialized at the beginning of each SPSS/PC+ session. The name of the log file can be changed at any time within a session by issuing a suitable SET command. The log from that point to the end of the session will then be stored on a file with the new name, and this will not be inititialized at the beginning of the next session. Log files may be edited with REVIEW within a session, and extracts from the edited file may be used as the basis of new command files. *See* Chapter 9.

SPSS MANAGER: a procedure used to remove SPSS/PC+ modules and to reinstall them to the system.

SPSS.PRC: the default name for a results file. This name can be changed by a suitable SET command, e.g.: SET RESULTS = "ATT.PRC".

SPSSPROF.INI: the file name given to the 'profile' file, a batch file containing commands which are automatically issued to the system each time SPSS/PC+ is entered from the particular directory. These commands often include SET commands which affect the operation of various parameters of the system (printer, beep, more and echo on/off status, for example, and log and listing file names). Once an SPSSPROF.INI file has been written and saved within a particular directory, the commands will be automatically executed each time the system is run from within that directory.

SPSS.SYS: the default name for a system file. A file of this name will be created if no file-name specification is given on a SAVE command, and this file will be read by a subsequent GET command which has no file-name specification. Only one 'SPSS.SYS' file can be present in any particular hard disk directory, so such a SAVE command will cause any existing file of that name to be overwritten by the new file. The default name can be changed by a suitable SET command.

STANDARD DEVIATION: a measure of the 'spread' or 'scatter' of values around the mean. If the frequency of values for a variable conforms to a normal distribution then knowledge of the mean and standard deviation allows an estimate to be made of the percentage of cases whose values fall below or above any particular value for that variable.

STANDARD ERROR: generally, a measure of the degree to which a statistic is likely to differ from its true value as a result of chance factors. For explanations of particular types of standard error (e.g. s.e. of skewness) refer to the entry for the relevant measure (e.g. skewness).

STANDARD SCORE: also known as a *Z*-score, this is a transformation of the raw score which takes into account the mean and standard deviation of the population. The standardized score is the difference between the raw score and the population mean, divided by the standard deviation. Standard scores have a mean of 0 and a standard deviation of 1 (although other transformations may be applied at a secondary stage). Such a transformation facilitates the comparison of results from different tests, for example.

STATISTICS: an SPSS/PC+ subcommand available for many procedures which allows the user to specify which statistical tests (in addition to the default tests for that procedure) should be performed.

STEM-AND-LEAF PLOT: a graphical display of a frequency distribution, rather like a histogram. However, whereas a histogram uses a single symbol (such as an asterisk) to represent all cases, a stem-and-leaf display provides the actual values for cases. Each row represents a stem (and is labelled with the leading digit(s) for the values plotted), and the final digit (the trailing digit) is printed within the row (the trailing digits form the leaves). Thus in the following example:

Frequency	Stem &	Leaf
5	0	24779
10	1	2445568889
7	2	1123357

the seven values represented in the final line of the display are 21, 21, 22, 23, 23, 25 and 27. A stem-and-plot display can be requested within an EXAMINE command.

STRATIFIED RANDOM SAMPLING: *See* stratified sampling.

STRATIFIED SAMPLING: a method of sampling in which a population is first divided into sub-groups, or 'strata', according to some characteristic, and a random sample then taken of each sub-group.

STRING VARIABLE: (also known as an 'alphanumeric variable'). A string is a sequence of characters. A string variable is one which is stored in character, rather than numerical, form. Because such variables are not numeric, only a limited number of operations (including sorting with the SORT command) can be performed with them, and they are not available for arithmetic operations. *See also* long string; short string.

SUBCOMMAND: an instruction which forms part of a command and specifies which operations are to be performed or the form in which output is to be displayed. For some commands, only, one or more subcommands are necessary. The most familiar subcommands are those which specify STATISTICS and OPTIONS.

SUBTITLE: a command used to provide a subtitle for each page of SPSS/PC+ output. *See* title; Chapter 16.

SUM: the result of adding each of the values for a particular variable in a set of data.

SURVEY: a technique for gathering data from or about a group of people or objects. Such methods include behavioural observation, written questionnaires and telephone or face-to-face interviews. In a sample survey the aim is to use a relatively small group (the sample) in order to derive conclusions or predictions about a larger group (the population). The extent to which valid extrapolations to the larger population will be possible will depend on the representativeness of the sample. *See*: sampling procedures.

SYSMIS: a special SPSS/PC+ function used in COMPUTE commands and conditional transformations and selections. The function produces the value 1 (or 'true') if the named variable has a system-missing value. Thus: SELECT IF (SYSMIS-(VAR101)). would retain in the active file only those cases for which VAR101 had a system-missing value.

SYSTEM FILE: a file created from an active file by using the SPSS/PC+ SAVE command. System files contain data, data definition information, system variables and other system information, and they are written in binary form. The use of system files can lead to considerable reduction in processing time, but they have the disadvantage that they cannot be edited.

SYSTEM-MISSING VALUE: a missing value automatically assigned by the SPSS/PC+ system in certain circumstances. Thus if a COMPUTE command specifies that C is to be created for each case by adding the value of A to the value of B, then if one or both of these values are missing from the original dataset then C, for that case, will be assigned a system-missing value. *See also* sysmis.

SYSTEM VARIABLE: the SPSS/PC+ system automatically assigns values to certain special variables for its own 'house-keeping'. These are known as 'system variables'. One such variable is *$casenum*, another is *$weight* (the names of system variables always begin with the $ symbol). Cases are assigned a *$casenum* value by the system as the file is made active. One way in which the user may employ this variable is to use it as a 'sorting key' (within a SORT command) to restore the original of cases within the active file: 'SORT BY $casenum'. The values of system variables can be listed using **LIST**, but the variable name(s) must be explicitly stated. *See also* $casenum; $weight.

TABLES :

(a) an optional module of the SPSS/PC+ system which produces high quality summary tables. *SPSS/PC+ Tables* provides the user with great flexibility in determining the appearance and content of tables which can include frequency counts, descriptive statistics and displays of multiple response data.

(b) a subcommand that may be used within a number of SPSS/PC+ commands, including CROSSTABS.

TABULATION: an arrangement of information in the form of a table. *See also* crosstabulation; contingency table.

TAIL: either extreme of a frequency distribution. The term is most commonly applied to probability distributions. A hypothesis about a difference between two sets of data can be stated in a directional form (e.g. boys will be taller than girls) or in a non-directional form (e.g. there will be a difference, one way or the other, in the heights of boys and girls). When a *t*-test is used to determine whether the means of two groups differ significantly the calculation will result in a particular t-value. To assess the significance of this value the directional or non-directional nature of the hypothesis must be taken into account. If the hypothesis is directional then a one-tail test applies. If the hypothesis is non-directional then a two-tail test applies.

TEST: a collection of items which can be scored together to yield one or several measurements of individual differences (e.g. personality, levels of ability). *See* scale.

TEST-RETEST RELIABILITY: reliability assessed by administering the same test twice to the same group of cases and then correlating the first set of scores with the second set of scores. *See* Chapter 14.

TEXT FILE: a file containing alphanumeric characters (and possibly some control characters). Text files are generally accessible for editing. Some files used by SPSS/PC+ are of this type although system files are not.

THRU: an SPSS/PC+ keyword used within the RECODE command to help specify a range of values to be recoded to a single new value. For example: RECODE VAR1 (1 THRU 99=1) (100 THRU 199=2). The range specified by THRU includes both of the limiting values, so that the expression '1 THRU 5' includes 1 and 5, and all values within this range.

TIED RANKS: when equal scores are ranked they are given tied ranks. There are a number of alternative ranking strategies, and tied ranks can be assigned in a number of different ways. The TIES subcommand within the RANK procedure specifies which of four methods is to be used. *See* Chapter 13.

TITLE: a command used to provide a page heading for SPSS/PC+ output. The default heading includes the date, the page number and 'SPSS/PC+'. If a TITLE has been specified then 'SPSS/PC+' is replaced with the user-specified title (up to 58 characters long). Thus: TITLE "Gender Questionnaire – Student Data". *See* Chapter 16.

TO: an SPSS/PC+ keyword used to list variables on a DATA LIST or PROCEDURE command. Thus 'VAR1 TO VAR44' would list 44 variables (VAR1, VAR2, VAR3 ... VAR43, VAR44). TO can also be used within the LIST command to refer to cases. Thus: LIST CASES FROM 1 TO 58. will produce a display of the values of all of the variables for cases 1 to 58. Use of the TO convention is not permitted when variables are listed on DROP or KEEP subcommands within a JOIN command.

TOGGLE: a switching between two states or modes of operation. In REVIEW, for example, pressing **<Ins>** repeatedly will cause the system to change between the 'insert' and 'overwrite' modes. This action thus operates the insert on/off toggle.

TRANSFORMATION: in SPSS/PC+, the act of changing the coding scheme for a variable (with a RECODE command) or of creating a new variable (with a COMPUTE command). SPSS/PC+ has powerful facilities for transforming data. *See also* conditional transformation; Chapter 8.

TRUNCATION: an SPSS/PC+ feature which allows the user to enter a keyword (e.g. CORRELATION, VARIABLES) in a shortened form. Only the first three letters are needed (e.g COR, VAR). Because of this feature, mis-spellings *after* the first three letters are ignored – thus CORELASHUN is an acceptable form of CORRELATION.

t-TEST: a parametric statistical test used to determine whether the means of two groups are significantly different. There are two versions of this test. The paired-samples *t*-test (also known as the matched pairs *t*-test, the related *t*-test, and the correlated *t*-test) compares the means of two paired sets of data. The two-sample *t*-test (also known as the independent *t*-test) compares the means of two unrelated or independent samples. *See* Chapter 7.

T-TEST: an SPSS/PC+ command used to request either a paired-samples or an independent-samples *t*-test. *See* Chapter 7.

TWO-TAIL TEST: a test for the significance of a difference between the means of two groups in a case in which the hypothesis has been cast in a non-directional form. This contrasts with a one-tail test. Using a two-tailed test, the null hypothesis is rejected if there is a small probability (the criterion is often .05 or less) of obtaining a result as extreme as the one observed, in either direction, when the null hypothesis is true. *See* null hypothesis; tail; significance tests.

TYPE I ERROR: making a judgement that an effect is present when in fact it is not. For example, concluding that there *is* a difference between two populations when, in fact, there is *not*.

TYPE II ERROR: making a judgement that an effect is not present when in fact it is. For example, concluding that two populations are *not* different when in fact they *are*.

UNEQUAL LENGTH SPEARMAN-BROWN COEFFICIENT: an estimate of the reliability of a whole test, based on a split-half reliability when the two halves are of unequal size. The formula used to derive this estimate assumes that the two halves of the test are equally reliable and have similar variance. The coefficient is included in the output from the RELIABILITY procedure when the split-half model is specified. *See also* split-half reliability; equal length Spearman-Brown coefficient; Guttman split-half coefficient. *See* Chapter 14.

USER-MISSING VALUE: a value for a particular variable which has been declared by the user (in a MISSING VALUE command) as representing data which are missing. The system recognizes the special nature of this value and, unless specifically requested to do so, will generally not treat the value literally. Thus if the value 9 has been declared as signifying missing data (e.g. for a variable for which only the values 1 to 8 are valid) any 9s encountered for this variable will not be included in calculations of averages, etc.). *See also* system-missing value.

VALID: in SPSS/PC+, any value for a variable other than that which has been declared by the user as a missing value and that which is a system-missing value.

VALIDITY: generally, the extent to which something meets a criterion of reasonableness or accuracy. Of a test or measure, the degree to which it measures what it is supposed to measure. There are several different ways of assessing the validity of a test. Typically, a set of scores provided by the test is compared with some external criterion, for example the performance in a task. A valid sample is one which has been appropriately selected and is therefore assumed to be representative of the larger population.

VALUE: a number or code which represents, for a particular case, the characteristic (size, degree, type, etc.) of the variable. The term is generally applied to quantified variables, but can also refer to those which have been coded into categories. The value in such cases refers to the code – a category *label* rather than a true number.

VALUE LABEL: an optional description of some or all of the particular values which a variable may take, e.g. (for *sex*) 1 "Male" 2 "Female". The output from many SPSS/PC+ procedures will include any relevant value labels which have been provided by the user. Value labels must always be enclosed in apostrophes (or quotation marks) when specified in a command. Although they may be up to 60 characters in length, it is preferable to keep them as succinct as possible. Value labels may be included within a data definition file or provided interactively during an SPSS/PC+ session.

VARIABILITY: the scatter of values in a sample.

VARIABLE: an item which can assume one of several different values or characteristics.

VARIABLE DEFINITION: that part of the DATA LIST command which declares the names of the variables represented by the data and specifies the format and position of each variable.

VARIABLE LABEL: a description of the variable which can be provided by the user, and is used by SPSS/PC+ to annotate output. Variable labels should always be enclosed in apostrophes (or quotation marks) when specified in commands, and although they may be up to 60 characters in length it is preferable to keep them succinct. Variable labels may be included within a data definition file or provided interactively during an SPSS/PC+ session.

VARIANCE: a measure of the dispersion of values. The square of the standard deviation.

WALD-WOLFOWITZ (TEST): a non-parametric test, also known as the Runs test, available within the NPAR procedure and used to determine whether a sequence of data with only two possible values (for example heads and tails) is random or contains runs of one or other value. *See* Chapter 13.

WARNING: the response made by the system to alert the user to a possible mistake. If a long string variable, for example, is specified in a command which applies only to numerical and short string variables then a warning is issued. The procedure continues, however, with the long string truncated to the first 8 characters (i.e. the system treats it as a short string). *See also* error.

WILCOXON (TEST): a non-parametric test, also known as the signed-ranks test, available within the NPAR procedure. It is used to determine whether matched pairs of data from two groups (at at least the ordinal level of measurement) show a significant difference. *See* Chapter 13.

$WEIGHT: a system variable which is automatically included for each case in the active file. The $ symbol at the beginning of the name indicates that the variable is a system variable rather than a user-declared variable. *$weight* is an extra variable for each case which represents a weighting to be applied to the case in analyses. The initial value of *$weight* for each case is 1.00. This can be changed by a WEIGHT command (e.g. WEIGHT BY VAR004.). The differential weighting can be cancelled with the command: WEIGHT OFF.

WILDCARD CHARACTER: a character which, when employed in certain DOS operations involving file-names, is interpreted to mean any single character or, alternatively, any group of characters. For example, when using DOS to copy files, the name of each file to be copied can be specified in full. Alternatively, the wildcard characters **?** and * can be employed. Thus the copying of three files named 'genone.dat', 'gentwo.def' and 'genthree.com' could be requested using a single command which includes the formulation 'copy gen?????.*'. Here each question mark would refer to any character and the asterisk would refer to any group of characters. The special formula '*.*' is used to copy *all* files (whatever their name and extension) from one location to another. When wildcard characters are used to copy a group of files each file retains its original name when copied to the second directory.

WINDOW: a part of the display screen which provides special information (for example, from within REVIEW, a menu or glossary definition). Several windows may be present at the same time.

WRITE:

(a) an SPSS/PC+ procedure which creates a file of cases together with variable names and format information. Records (lines) are 80 characters in length and are formatted with a single blank space between variables.

(b) an SPSS/PC+ subcommand available for several procedures, requesting that an output file be written to disk. Results files are also written if the user specifies certain OPTIONS within a number of procedure commands (including CORRELATION and ONEWAY).

X-AXIS: the horizontal axis (abscissa) of a graph.

Y-AXIS: the vertical axis (ordinate) of a graph.

YATES' CORRECTION: a correction factor used in a chi-square test when the design has only one degree of freedom. Also referred to as the correction for continuity.

Z SCORE: a standardized score.

Appendix H: Types of File Used or Created by SPSS/PC+

File type	Includes	Written by USER(U) or SYSTEM(S)	Write Command	Default Name	Read Command	Type (EXT.)	Edit?
Data	Data	U	(Editor)	None	DATA LIST	ASCII (.DAT)	Yes
Data definition	Data List, labels, etc. (and can also include data)	U	(Editor)	None	INCLUDE	ASCII (.DEF)	Yes
Command	Commands (and can also include data definition info. and data).	U	(Editor)	None	INCLUDE	ASCII (.SPS)	Yes
System	Data and definition information	S	SAVE	SPSS.SYS	GET	binary (.SYS)	No
Listing	Output from any procedures during current session	S	None (auto)	SPSS.LIS	REVIEW	ASCII (.LIS)	Yes
Log	A record of input during current session	S	None (auto)	SPSS.LOG	REVIEW	ASCII (.LOG)	Yes
Portable	As system files (+ some system information)	S	EXPORT	None	IMPORT	ASCII (.POR)	Not easily
Results	Matrix of summary results	S	approp. OPTION selected	SPSS.PRC	INCLUDE (or DATA LIST)	ASCII (.PRC)	Yes

Appendix I: Variations Between Different Versions of SPSS/PC+

The version of SPSS currently used on mainframe computers is known as $SPSS^X$ and the first microcomputer version of the system was SPSS/PC. That was superseded by SPSS/PC+ which is, at the time of writing, available in Version 4.0.

The different versions overlap to a large extent, with later editions generally adding new procedures and expanding the capabilities of existing procedures.

Those who are working with earlier versions of SPSS/PC+ may wish to know how their system differs from the one discussed in this book (Version 4.0).

The following is a brief history of the development of SPSS/PC+ as it has evolved between the original version and Version 4.

SPSS/PC+ V2

REVIEW: Considerably enhanced, and changed in many ways:

- many commands changed
- new menu and help system
- menu of variable names
- on-line glossary available.

The command used to run REVIEW from **DOS** was modified, to become: (e.g.):

C>spsspc/re file.def

New base system commands: RELIABILITY, MODIFY VARS, SYSFILE INFO.

SPSS/PC+ V3

The number of variables possible in an active file or a system file was increased from 200 to 500.

REVIEW: Several additional features were included to add to the efficiency of editing.

New base system commands: RANK, EXAMINE, AUTORECODE, FLIP.

Other new procedures: NLR and LOGLINEAR added to *SPSS/PC+ Advanced Statistics* module.

SPSS/PC+ V4

Major reorganisation of modules. Some procedures that were formerly part of the BASE system were moved into the new *SPSS/PC+ Statistics* module, together with some procedures formerly included in the *SPSS/PC+ Advanced Statistics* module.

Additional features were added to FREQUENCIES, CORSSTABS, GET and SAVE.

New base system command: ADD VALUE LABELS

Other new procedures SURVIVAL and PROBIT added to *SPSS/PC+ Advanced Statistics* module.

Appendix J: Brief Guide to SPSS/PC+ Commands and Subcommands

If you use SPSS/PC+ sporadically, you may need to be reminded of even the most basic rules of syntax of the various commands. The following examples of commands will serve to remind you that when using the **CROSSTABS** procedure the syntax is 'x BY y' whereas when using the **CORRELATION** procedure the syntax is 'x WITH y', etc. The guide also indicates some of the most frequently used subcommand options.

THIS GUIDE IS NOT COMPREHENSIVE AND SHOULD BE USED ONLY AS A MEMORY AID AFTER YOU HAVE WORKED THROUGH THE RELEVANT CHAPTERS OF THIS BOOK

ANOVA (Performs analysis of variance)

```
ANOVA VAR1 VAR3 TO VAR5 BY AGEGROUP(1,4) VAR8 (1,3).
```

This will produce four two-way anovas. Each of the dependent variables (*var1*, *var3*, *var4* and *var5*) will be subjected to separate analysis. In each analysis there are four levels of the independent integer variable *agegroup* and three levels of the second independent integer variable *var8*.

Options: 2. Suppress labelling
9. Regression approach
Statistics: 3. Display counts and cell means

COMPUTE (Creates a new variable, producing values for each case)

```
COMPUTE VAR101=VAR1 + VAR2 + VAR3.
COMPUTE VAR102=VAR1 + (VAR2/VAR3)**2.
```

CORRELATION (Performs Pearson product-moment correlations)

```
VAR1 VAR2 VAR3.
```

This will produce 3 correlations (1–2, 1–3, 2–3)

```
CORRELATION VAR1 TO VAR10 WITH VAR2 TO VAR10.
```

This will produce a matrix of 90 correlations, including nine in which a variable is correlated with itself (the value therefore will be 1.000)

Options:	2. Pairwise deletion
	3. Two-tailed probability
	4. Write matrix results file
	5. Display count and probability values
Statistics:	1. Display of variable counts, means and standard deviations

(*See also*: rank correlation.)

CROSSTABS (Crosstabulates data on two or more integer variables)

```
CROSSTABS VAR1 VAR2 VAR3 BY VAR4 VAR8.
```

This will produce six cross-tabulation tables (1–4, 1–8, 2–4, 2–8, 3–4, 3–8). The first named variables (*var1*, *var2*, *var3*) will form the rows of the tables, the second named variables (*var4*, *var8*) the columns.

Cells:	ROW COLUMN TOTAL EXPECTED, etc. (also NONE and ALL)
Statistics:	CHISQ PHI GAMMA CC & 9 others (also NONE or ALL)

DATA LIST (fixed format) (Lists variables and indicates their position within the data set)

```
DATA LIST FILE= "GENDER.DAT" /VAR1 1-2 VAR2 3-5 VAR3 6 VAR4 7.
```

For data to be read from an external matrix file "gender.mat" :

```
DATA LIST MATRIX FILE= "GENDER.MAT" / V1 1-2 V2 3-5 V3 6.
```

DISPLAY (Provides information about variables in the active file)

a) DISPLAY.
b) DISPLAY ALL.
c) DISPLAY VAR8 TO VAR14.

b) and c) will provide fuller information about variables than will a).

EXPORT (Produces a portable ASCII version of the active file)

```
EXPORT OUTFILE= "GENDER.POR".
```

(produces a portable file "gender.por" containing all of the current active file – to produce a file of only part of the contents use Drop and/or Keep subcommands)

```
EXPORT OUTFILE= "GENDER.POR"/DROP = VAR15 TO VAR19.
```

The TO convention *is* permitted on the Drop or Keep subcommand.

FREQUENCIES (Displays frequency counts for values of single variables)

```
FREQUENCIES VAR1 TO VAR26.
```

Barchart: displays a barchart (parameters can be specified)

Histogram: displays a barchart (parameters can be specified)

Statistics: The default statistics are : MEAN STDDEV MINIMUM and MAXIMUM. Others which may be requested include RANGE MODE MEDIAN VARIANCE. The user may also request ALL (or NONE):

```
/STATISTICS = DEFAULT RANGE VARIANCE.
```

(this would request the default statistics plus the two others).

GET (Reads a system file)

```
GET FILE= "GENDER.SYS".
```

(Drop and Keep subcommands may also be included, and for these the TO convention *is* permitted).

IF (Transforms or creates a variable if a logical condition is fulfilled)

```
IF (VAR1 GT 3) GROUP=2.
IF (VAR3 NE VAR4) VAR101=2.
IF (VAR3 EQ 1 OR VAR3 EQ 2 OR VAR3 EQ 3) VAR58 = VAR3+ 12-VAR6.
```

In the last of these examples, if the value of *var3* for a case equals 1, 2 or 3 then (and only then) a variable *var58* for that case is computed by adding the value 12 to the *var3* value and then subtracting the value of *var6*. If *var3* for the case does not equal 1, 2 or 3 then *var58* for that case retains its original value or is 'system-missing'.

IMPORT (Reads a portable ASCII file containing data and data dictionary)

```
IMPORT FILE = "GENDER.POR".
```

(Drop or Keep, Rename and Map subcommands may also be included).
The TO convention *is* permitted on a Drop or Keep subcommand.

INCLUDE (Used to read a command file, include a file which itself contains an INCLUDE command – such 'nesting' can be five levels deep)

```
INCLUDE "GENDER.DEF".
INCLUDE "TEST.SPS".
```

JOIN (Combines up to five files – one can be the active file, all others must be system files)

With the keyword MATCH, a JOIN command combines different variables for the same cases :

```
JOIN MATCH FILE= "GENPRE.SYS"
/FILE= "GENPOST.SYS".
```

With the keyword ADD, a JOIN command combines different cases which have data for the same variables :

```
JOIN ADD FILE= "GENMALE.SYS"
/FILE= "GENFEM.SYS".
```

With either MATCH or ADD, the subcommands Drop or Keep, Rename, By and Map may also be included. If no By subcommand is included in a JOIN MATCH command then cases are matched in the order in which they occur in the two files (first with first, second with second, etc.). If By is included it must be placed as the last subcommand (except for Map) and the 'key variable' used to match the cases (in the following example it is *id*) must be present in all files to be matched. The files must also be in prior order by the key variable (thus it might be necessary to SORT them before the JOIN). A By subcommand with ADD can be used to determine the order in which cases will occur in the new file. The current active file must be indicated by an asterisk :

```
JOIN MATCH FILE = *
/DROP VAR34 VAR38 VAR44
/FILE= "GENPOST.SYS"
/DROP VAR134 VAR138
/BY ID
/MAP.
```

The TO convention is *not* permitted on a Drop or Keep subcommand

---

LIST (Displays the values of all (or specified) variables for cases in the active file).

```
LIST ALL.
LIST VAR1 TO VAR16.
```

Data on specific cases only can also be requested :

```
LIST ID VAR1 TO VAR16
/CASES= FROM 1 to 20.
```

MEANS (Displays means, standard deviations and group counts for dependent variables defined by independent variable(s))

```
MEANS VAR1 VAR2 VAR4 BY SEX AGEGROUP.
```

var1, *var2* and *var3* are the dependent variables; *sex* and *agegroup* are the independent variables. Six tables are requested. Compare with :

```
MEANS VAR1 VAR2 VAR4 BY SEX BY AGEGROUP.
```

Groupings by *sex* are made and, within these, *agegroup* subgroups. Thus mean values will be given for 'old males', 'young females', etc. This command requests three tables.

Options: 3. Suppress all labels 12. Display group variances
Statistics: 1. One-way analysis of variance

MISSING VALUE (Declares a value indicating that a datum for that variable is missing for that case).

```
MISSING VALUE VAR1 VAR4 TO VAR20 (9) VAR2 VAR3 (99).
```

N (A case selection command. Limits the number of cases in the 'working file' to the first N cases)

```
N 50.
```

NPAR TESTS (Performs a number of non-parametric tests)

```
NPAR TESTS KENDALL = VAR1 VAR2 VAR4 VAR6.
```

(*See* Chapter 13 for details of each test)

Options: 2. Listwise deletion 3. Sequential pairing of data from two related groups

Statistics: 1. Count, mean, standard deviation, minimum and maximum
2. Quartiles and count

ONEWAY (Performs one-way analysis of variance)

```
ONEWAY VAR1 VAR3 VAR5 BY AGEGROUP(1,3).
```

A number of 'post-hoc' multiple contrast tests can be requested by the Ranges subcommand:

```
ONEWAY VAR1 BY AGEGROUP(1,3)
/RANGES=DUNCAN
```

Options: 2. Listwise deletion 4. Write matrix
7. Read matrix

The matrices referred to in Options 4 and 7 include counts, means and standard deviations.

Statistics: 1. Group descriptive statistics

PLOT (Produces two dimensional graphical plots of bivariate distributions)

```
PLOT PLOT VAR4 WITH VAR7.
```

PLOT subcommands include:

Title – provides heading
Vertical – provides label for vertical axis
Horizontal – provides label for horizontal axis
Format – specifies the type of plot to be produced

PROCESS IF (Produces a temporary selection of cases to be included in the next procedure)

PROCESS IF selects cases according to the value of each case on a specified variable.

```
PROCESS IF (VAR3 GE 3).
PROCESS IF (VAR3 LT 2).
```

N.B. Logical operators (AND, OR, NOT) cannot be included in PROCESS IF commands.

The following relational operators are permitted :

EQ (or =)	**GT** (or >)	**LT** (or <)
NE (or <>)	**GE** (or >=)	**LE** (or <=)

RANK (ranks variables, creating for each one a new variable, the values of which correspond to either an ascending or a descending rank-ordering of the original variable. Also used to create percentile value variables).

```
RANK SALARY (D).
```

The new variable *rsalary* is created for each case. The values of the new variable reflect, in the example, the descending ranking of the original *salary* variable for that case.
New variable given the label "RANK of (ORIGINAL)".

Rank Into: used to specify name for new variable:

```
RANK SALARY / RANK INTO RANKSAL.
```

By: used to rank within groups

```
RANK SALARY (D) BY SEX.
```

Ties: four alternative strategies available for dealing with tied ranks: `Mean; Low; High; or Condense`

```
RANK SEXISM(D) / TIES = LOW.
```

Default strategy is Mean.

Ntiles: creates a new variable by grouping ranked values into a specified number of percentile groups. The new variable has values from 1 to *n*, where *n* is the number of specified groups. Thus:

```
RANK HT / NTILES(4).
```

divides cases into quartiles on the basis of values for *ht* (new variable, with values 1 to 4, would be *nht*, unless Into used:

```
RANK HT/ NTILES(4) INTO QUARTHT.
```

Rank-order correlation

Use RANK for variables, then use CORRELATION specifying the new (rank) variables to obtain a Spearman Rank Correlation Coefficient.

```
RANK   SEXISM  NEUR.
CORRELATION  RSEXISM RNEUR.
```

RECODE (Revises the coding scheme for an existing variable)

```
RECODE VAR1 (1,2,3=1) (4=2) (5,6,7=3).
```

Permitted keywords include : LO, HI, THRU and ELSE :

```
RECODE SALARY (LO THRU 999=1) (1000 THRU 1999=2) (ELSE=3).
```

The keywords HI, LO and ELSE should be used with extreme care to avoid the inadvertent recoding of missing values as valid.

RELIABILITY: (Assesses reliability of an additive multi-variable scale and examines relationships between the scale and individual items).

```
RELIABILITY / VARIABLES = VAR1 VAR2 VAR6 VAR10 VAR18
/ MODEL = SPLIT
/ SCALE (RACISM) = ALL / STATISTICS = DESC CORRELATIONS SCALE
/ SUMMARY = MEANS CORRELATIONS TOTAL.
```

Model: Five different types of model available, including:

ALPHA – the internal consistency model
SPLIT – the 'split-half' model

Alpha is the default model used if the Model subcommand is omitted.

Scale: Specifies scale name and variables to be included

Statistics: DESC – means and standard deviations for each item.
CORRELATIONS – inter-item correlation matrix
SCALE – mean and variance for the scale(s)
Six other statistics specifications available.

Summary: requests summary statistics for scale items.

MEANS – takes item means and gives average, minimum value, maximum value, range, ratio of minimum to maximum, and variance of the means.
VARIANCE – same summary statistics, but on item variances
CORRELATIONS – same summary statistics, but for correlations between every pair of scale items.
TOTAL – statistics comparing each item with scale composed of all *other* items, including scale mean, variance, and Cronbach's alpha. Also correlation between each item and the scale composed of all other items.

Although working with a scale, the command does not add a new (scale) variable to the active file (this can be done following a RELIABILITY analysis, by using an appropriate COMPUTE command).

REVIEW (Allows the editing of text files, including data files, data definition files, the log file and the listing file)

See *Appendix K* for command summary.

SAMPLE (Draws temporary random selection of cases to be processed in the next procedure)

```
SAMPLE .25.
SAMPLE 100 from 253.
```

SAVE (Saves current active file as a system file)

```
SAVE.
```

Produces system file with default name SPSS.SYS (unless default has been changed)

```
SAVE OUTFILE = "GENDER.SYS"
```

Produces system file named "gender.sys"

N.B. System files cannot be edited.

(a Drop subcommand may also be included – but there is *no* Keep subcommand. The TO convention *is* permitted within the Drop subcommand).

SELECT IF (Selects – according to a logical criterion – certain cases to remain in the active file. The selection remains in effect until the end of the session or until another file is made active)

SELECT IF selects cases according to the value of each case on a specified variable.

```
SELECT IF (VAR3 GE 3 OR VAR4 GE 3).
SELECT IF (VAR3 LT 2).
SELECT IF (VAR3 GT (VAR4*10)).
```

Thus relational operators (EQ, GE, NE, etc.) and logical operators (AND, OR, NOT) can be included in SELECT IF commands.

The logical expression is evaluated. If true, the case is selected. If false, the case is not selected.

SET (Changes any of a wide range of optional features of the SPSS/PC+ operation)

Use SHOW command to display current settings.

```
SET PRINTER = ON.
SET MORE = OFF.
SET LISTING = "GEND.LIS".
SET PRINTER ON / LOG= "GEND.LOG" / BEEP OFF.
```

SHOW (Displays current settings for a wide range of optional features of the SPSS/PC+ operation)

```
SHOW.
```

To change any of the features displayed use the SET command (see above).

SORT (Re-orders cases in the active file according to the value of each case on one or more – up to 10 – 'sort keys')

`SORT BY AGE.`	Sorts cases in ascending order of age
`SORT BY AGE(D).`	Sorts cases in descending order of age
`SORT BY AGE(A) SCORE(A) HT(D)`	Sorts cases first by age – in ascending order. Cases in the same age category are further sorted by score, also in ascending order. Any cases with the same age and score are sorted by descending order of *ht*.

SUBTITLE (*See* **TITLE**)

TITLE and **SUBTITLE** (Changes default heading and adds subtitle to pages of output)

By default, each page of SPSS/PC+ output includes the date, a page number and a centre title 'SPSS/PC+'. This title is changed if the user issues a TITLE command followed by up to 58 characters followed by a terminator (.). The use of quotation marks will allow printing of both upper and lower case and the inclusion of apostrophes, etc.

```
TITLE "First stage of Michael's Correlational Analysis".
```

Similarly, SUBTITLE causes a user-supplied label to be included beneath the title line:

```
SUBTITLE "Male Subjects' Data".
```

Subtitles are not centred but start in the first column. To cancel a title or subtitle enter the single command word only.

T-TEST (Performs an independent *t*-test or paired *t*-test to test the difference between the means of two samples)

```
T-TEST GROUPS= SEX(1,2)/ VARIABLES = VAR1 TO VAR12.
```

This command would be used to perform 12 **independent** *t*-tests on the two groups coded 1 and 2 for the variable *sex*.

Output includes *t*-values and probabilities based on both **pooled variance** and **separate variance**. The separate variance estimate should generally be used if the *F*-value provided in the table is associated with a probability of less than .05. Otherwise the *t*-value and probability based on the pooled variance estimate should be used.

```
T-TEST PAIRS= VAR1 VAR2 VAR3 VAR4.
```

This command would be used to perform six **paired** *t*-tests; the values of all valid cases in the active file would be compared (in pairwise fashion) on each possible pair of variables in the list.

```
T-TEST PAIRS= VAR1 VAR2 WITH VAR3 VAR4.
```

This command would be used to perform four **paired** *t*-tests; each of *var1* and *var2* would be compared with each of *var3* and *var4*.

Options:	2. Listwise deletion 3. Variable labels suppressed 5. WITH interpreted to execute 'sequential' pairing
Statistics:	No 'additional' statistics are available. The default display includes *t*-value, degrees of freedom, and two-tailed probabilities. Counts, means, standard deviations and standard error are calculated for each variable or group. Other statistics are also provided.

VALUE LABELS (Provides labels for some or all of the values which a variable may take)

```
VALUE LABELS SEX 1 "Female" 2 "Male" /AGE 1 "Under 20"
2 "20+" / VAR37 TO VAR41 1 "YES" 2 "NOT SURE" 3 "NO"
/VAR23 VAR25 VAR27 1 "Strongly Agree" 7 "Strongly Disagree".
```

VARIABLE LABELS (Provides labels for variables)

```
VARIABLE LABELS VAR1 TO VAR6 "Attitudes to men"
/VAR7 TO VAR12 "Attitudes to women".
```

Appendix K: Brief Guide to Review Commands

To enter review directly from DOS:

```
C>SPSSPC/RE  filename    <RETURN>
```

Command	**Key(s)**
Switch 'insert' on/off	**<Ins>**
F1 – INFORMATION	
REVIEW help	**<F1><r>**
SPSS/PC+ help	**<F1><h>**
Files list	**<F1><f>**
Variables list	**<F1><v>**
Glossary	**<F1><g>**
Switch menu and help system on	**<F1><m>**
F3 – FILES	
To insert a file	**<F3><i>**
To edit another file	**<F3><e>**
F4 – LINES	
Delete a line	**<F4><d>**
Undelete a line	**<F4><u>**
Insert a line after current cursor position	**<F4><i>**
Insert a line before current cursor position	**<F4><b>**
F5 – STRING SEARCH AND CHANGE	
*F*orward search	**<F5><f>**
*B*ackward search	**<F5><b>**
F*O*rward change	**<F5><o>**
B*A*ckward change	**<F5><a>**

Command	**Key(s)**
F7 & F8 – AREAS	
Mark area of lines	**<F7><l>**
then move cursor to end line and	**<F7>**
Mark rectangle area	**<F7><r>**
then move cursor to end line and	**<F7>**
Unmark area	**<F7>**
Delete area	**<F8><d>**
Copy area	**<F8><c>**
Move area	**<F8><m>**
F9 – SAVE	
Write a whole file to disk	**<F9><w>**
Write marked area to disk	**<F9><m>**
EXIT from **REVIEW**	**<Alt-F10>**

REVIEW ESCAPES

To leave **REVIEW**	**<Alt-F10>**
To climb one level of menu system	**<Esc>**
To climb to top (Main Menu)	**<Alt-Esc>**
To cancel menus	**<Alt-m>**
To cancel prompt (for file name, etc.)	**<Esc>**
To return to editing after using Glossary	**<Esc>**
To remove information, etc.	Move cursor

Index

Entries in upper case are SPSS/PC+ commands, or other key words. Entries which are included in the Glossary (*Appendix G*) are marked **G**.